Student Solutions Manual for Tussy and Gustafson's
Prealgebra

Catherine Gong
Citrus College

Brooks/Cole Publishing Company

I(T)P® *An International Thomson Publishing Company*

Pacific Grove • Albany • Belmont • Bonn • Boston • Cincinnati • Detroit
Johannesburg • London • Madrid • Melbourne • Mexico City • New York
Paris • Singapore • Tokyo • Toronto • Washington

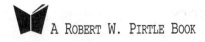
A Robert W. Pirtle Book

Sponsoring Editor: *Beth Wilbur*
Editorial Associate: *Nancy Conti*
Production: *Dorothy Bell*

Cover Design: *Vernon T. Boes*
Cover Photo: *Pierre-Yves Goavec/Image Bank*
Printing and Binding: *Patterson Printing*

COPYRIGHT © 1997 by Brooks/Cole Publishing Company
A division of International Thomson Publishing Inc.
I(T)P The ITP logo is a registered trademark under license.

For more information, contact:

BROOKS/COLE PUBLISHING COMPANY
511 Forest Lodge Rd.
Pacific Grove, CA 93950
USA

International Thomson Editores
Seneca 53
Col. Polanco
11560 México, D. F., México

International Thomson Publishing Europe
Berkshire House 168-173
High Holborn
London WC1V 7AA
England

International Thomson Publishing GmbH
Königswinterer Strasse 418
53227 Bonn
Germany

Thomas Nelson Australia
102 Dodds Street
South Melbourne, 3205
Victoria, Australia

International Thomson Publishing Asia
221 Henderson Road
#05-10 Henderson Building
Singapore 0315

Nelson Canada
1120 Birchmount Road
Scarborough, Ontario
Canada M1K 5G4

International Thomson Publishing Japan
Hirakawacho Kyowa Building, 3F
2-2-1 Hirakawacho
Chiyoda-ku, Tokyo 102
Japan

Printed in the United States of America

10 9 8 7 6 5 4 3 2

ISBN 0-534-34386-4

Table of Contents

STUDY SET Section 1.1

VOCABULARY

1. A <u>set</u> is a collection of objects.

3. The numbers 1, 2, 3, 4, 5, . . . form the set of <u>natural numbers</u>.

5. When 297 is written as 2 hundreds + 9 tens + 7 ones, it is written in <u>expanded</u> <u>notation</u>.

CONCEPTS

7. A 3 is in the tens column.

9. A 6 is in the hundreds column.

11. The set of whole numbers is obtained.

NOTATION

13. The symbol > means <u>is greater than</u>.

PRACTICE

15. $61 went for losses involving drunk drivers.

17. Emmitt Smith had 22 rushing touchdowns.

19. The closing value was 5177.

21.

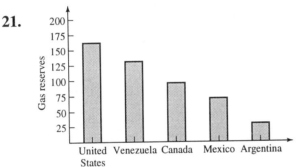

23.

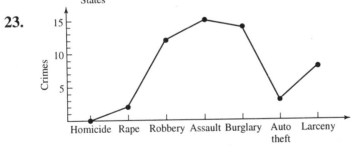

25. 2 hundreds + 4 tens + 5 ones; two hundred forty-five

27. 3 thousands + 6 hundreds + 9 ones; three thousand six hundred nine

29. 3 ten thousands + 2 thousands + 5 hundreds; thirty-two thousand five hundred

31. 1 hundred thousand + 4 thousands + 4 hundreds + 1 one; one hundred four thousand four hundred one

33. 425

35. 2,736

37. 456

39. 27,598

41. 660

43. 138

45. 863

47. 79,590

49. 80,000

51. 5,926,000

53. 5,900,000

55. $419,160

57. $419,000

APPLICATIONS

59. Fifteen thousand six hundred one and $\frac{\text{no}}{100}$

61. a. This diploma awarded this twenty-seventh day of June, one thousand nine hundred and ninety-six

    b. The suggested contribution is eight hundred fifty dollars a plate or an entire table may be purchased for five thousand two hundred and fifty dollars.

63. 299,800,000 m/sec

STUDY SET Section 1.2
VOCABULARY

1. Numbers that are to be added are called <u>addends</u>.

3. A <u>variable</u> is a letter that stands for a number.

5. A <u>square</u> is a rectangle with all sides of equal lengths.

7. The property that guarantees that we can add two numbers in either order and get the same sum is called the <u>commutative</u> property of addition.

9. The distance around a rectangle is called its <u>perimeter</u>.

CONCEPTS

11. commutative property of addition

13. associative property of addition

15. $x + y = y + x$

17. Any number added to <u>0</u> stays the same.

NOTATION

19. The symbols ( ) are called <u>parentheses</u>.

21. 33 plus 12 equals 45.

23. $(36 - 11) + 5 = \underline{\ 25\ } + 5$

PRACTICE

25. $25 + 13 = 38$

27. $156 + 305 = 461$

29. $(95+16)+39 = 111+39$
$$= 150$$

31. $25+(321+17) = 25+338$
$$= 363$$

33.
$$\begin{array}{r} 632 \\ +347 \\ \hline 979 \end{array}$$

35.
$$\begin{array}{r} 1,372 \\ +613 \\ \hline 1,985 \end{array}$$

37.
$$\begin{array}{r} 6,427 \\ +3,573 \\ \hline 10,000 \end{array}$$

39.
$$\begin{array}{r} 8,539 \\ +7,368 \\ \hline 15,907 \end{array}$$

41.
$$\begin{array}{r} 1,246 \\ 578 \\ +37 \\ \hline 1,861 \end{array}$$

43.
$$\begin{array}{r} 3,156 \\ 1,578 \\ +578 \\ \hline 5,312 \end{array}$$

45.
$$\begin{array}{r} 12 \\ 12 \\ 32 \\ +32 \\ \hline 88 \end{array}$$
The perimeter is 88 ft.

47.
$$\begin{array}{r} 17 \\ 17 \\ 17 \\ +17 \\ \hline 68 \end{array}$$
The perimeter is 68 in.

49. $17-14 = 3$

51. $39-14 = 25$

53. $174-71 = 103$

55. $633-598-30 = 35-30$
$$= 5$$

57.
$$\begin{array}{r} 367 \\ -343 \\ \hline 24 \end{array}$$

59.
$$\begin{array}{r} 423 \\ -305 \\ \hline 118 \end{array}$$

61.
$$\begin{array}{r} 1,537 \\ -579 \\ \hline 958 \end{array}$$

63.
$$\begin{array}{r} 4,267 \\ -2,578 \\ \hline 1,689 \end{array}$$

65.
$$\begin{array}{r} 17,246 \\ -6,789 \\ \hline 10,457 \end{array}$$

67.
$$\begin{array}{r} 15,700 \\ -15,397 \\ \hline 303 \end{array}$$

APPLICATIONS

69. $27-21 = 6$
She had 6 seedlings left.

71. $23-5 = 18$
The fare was $18.

73. $(370+40)-197 = 410-197$
$$= 213$$
The account balance is $213.

75.

$$
\begin{array}{r}
2,345 \\
1,712 \\
1,778 \\
445 \\
1,003 \\
+\ 2,774 \\
\hline
10,057
\end{array}
$$

She drove a total of 10,057 miles

77.

$$
\begin{array}{r}
18 \\
20 \\
+\ 24 \\
\hline
62
\end{array}
$$

The length of the house is 62 ft.

79. $34 + 34 + 64 + 64 = 196$
The flag needs 196 inches of fringe.

81. $18,549 + 25,182 = 43,731$
There were 43,731 graduates.

83. $9 + 26 = 35$
He was in space for 35 hours.

85. $5,305 - 5,272 = 33$
The Dow rose 33 points.

REVIEW

91. 3 thousands + 1 hundreds + 2 tens + 5 ones

93. 6,354,780

95. 6,350,000

STUDY SET Section 1.3

VOCABULARY

1. Multiplication means repeated addition.

3. A product is the result of a multiplication problem.

5. The statement $(a \bullet b) \bullet c = a \bullet (b \bullet c)$ expresses the associative property of multiplication.

7. The result of a division problem is called a quotient.

CONCEPTS

9. $4 \bullet 8$

11. $(xy)z = x(yz)$

13. A number can never be divided by 0.

NOTATION

15. $\times , \bullet , (\ )$

PRACTICE

17. $4 \bullet 7 = \underline{28}$

19. $12 \bullet 7 = \underline{84}$

21. $27 \bullet 12 = \underline{324}$

23. $9 \bullet (4 \bullet 5) = \underline{180}$

25.

$$
\begin{array}{r}
99 \\
\times\ 77 \\
\hline
693 \\
693 \\
\hline
7,623
\end{array}
$$

27.

$$
\begin{array}{r}
20 \\
\times\ 53 \\
\hline
60 \\
100 \\
\hline
1,060
\end{array}
$$

29.
$$\begin{array}{r} 112 \\ \times\ 23 \\ \hline 336 \\ 224 \\ \hline 2{,}576 \end{array}$$

31.
$$\begin{array}{r} 207 \\ \times\ \ 97 \\ \hline 1449 \\ 1863 \\ \hline 20{,}079 \end{array}$$

33. $13{,}456 \cdot 217 = 2{,}919{,}952$

35. $3{,}302 \cdot 15{,}358 = 50{,}712{,}116$

37. $12 \cdot 11 = 132$
She earned \$132.

39. $14 \cdot 29 = 406$
The car can go 406 miles on one tank.

41. $2 \cdot 37 \cdot 1{,}700 = 125{,}800$
About 125,800 fans heard the group.

43. $13 \cdot 24 = 312$
It takes 312 oranges for a case of 24 cans.

45. $14 \cdot 150 = 2{,}100$
Yes – there would be 2,100 pounds, which is greater than the maximum capacity.

47. $A = l \cdot w$
$ = 14 \cdot 6$
$ = 84$
The area is 84 in.^{2}.

49. $A = l \cdot w$
$ = 12 \cdot 12$
$ = 144$
The area is 144 in.^{2}.

51. $40 \div 5 = \underline{8}$

53. $42 \div 14 = \underline{3}$

55. $132 \div 11 = \underline{12}$

57. $\dfrac{221}{17} = \underline{13}$

59.
$$\begin{array}{r} 73 \\ 13\overline{)949} \\ \underline{91} \\ 39 \\ \underline{39} \\ 0 \end{array}$$

61.
$$\begin{array}{r} 41 \\ 33\overline{)1{,}353} \\ \underline{132} \\ 33 \\ \underline{33} \\ 0 \end{array}$$

63.
$$\begin{array}{r} 205 \\ 39\overline{)7{,}995} \end{array}$$

65.
$$\begin{array}{r} 210 \\ 29\overline{)6{,}090} \end{array}$$

67.
$$\begin{array}{r} 8 \\ 31\overline{)273} \\ \underline{248} \\ 25 \end{array}$$

quotient = 8;
remainder = 25

69.
$$\begin{array}{r} 20 \\ 37\overline{)743} \\ \underline{74} \\ 03 \end{array}$$

quotient = 20;
remainder = 3

71.

$$\begin{array}{r} 30 \\ 42\overline{)1{,}273} \\ \underline{126} \\ 13 \end{array}$$

quotient = 30;
remainder = 13

73.

$$\begin{array}{r} 31 \\ 57\overline{)1{,}795} \\ \underline{171} \\ 85 \\ \underline{57} \\ 28 \end{array}$$

quotient = 31;
remainder = 28

APPLICATIONS

75. $73 \div 23 = 3$, remainder = 4
4 cartons were left.

77. $950{,}000 \div 16 = 59{,}375$
59,375 gallons are drained each hour.

79. $P = 2 \cdot 360 + 2 \cdot 270$
$ = 720 + 540$
$ = 1260$

The perimeter is 1260 mi.
$A = 360 \cdot 270$
$ = 97{,}200$

The area is 97,200 mi.^{2}.

81. $8 \cdot 8 = 64$
There are 64 squares on a chessboard.

83. $A = 27 \cdot 19$
$ = 513$
There are 388 ft^2 left for planting.

85. $2 \cdot 5{,}280 = 10{,}560$
$11{,}000 - 10{,}560 = 440$
11,000 ft is 440 ft more than 2 miles.

87. singles:
$$A = 27 \cdot \frac{1}{2} \cdot 78$$
$$= 1{,}053$$
doubles:
$$A = \frac{1}{2} \cdot 36 \cdot \frac{1}{2} \cdot 78$$
$$= 702$$

$1{,}053 - 702 = 351$

The singles player must defend 1,053 ft^2; the doubles player must defend 702 ft^2.

The difference is 351 ft^2.

89. $954{,}193 \div 23{,}273 = 41$
Each book costs $41.

REVIEW

95. 8 is in the hundreds column.

97. 46,000

99. $357 + 39 + 476 = 872$

101. $97 - 75 = 22$

STUDY SET Estimation: Whole Numbers

1.

$$\begin{array}{r} 30{,}000 \\ 10{,}000 \\ 9{,}000 \\ 1{,}000 \\ 10{,}000 \\ + 30{,}000 \\ \hline 90{,}000 \end{array}$$

3.

$$\begin{array}{r} 500 \\ \times 70 \\ \hline 35{,}000 \end{array}$$
No, not reasonable.

5. $60{,}000 \div 30 = 2{,}000$
No, not reasonable.

7.

$$
\begin{array}{r}
4,000 \\
600 \\
1,000 \\
300 \\
+3,000 \\
\hline
8,900
\end{array}
$$

She flew approximately 8,900 miles.

9. $90,000 \div 3,000 = 30$
About 30 bags are needed.

STUDY SET Section 1.4

VOCABULARY

1. Numbers that are multiplied together are called <u>factors</u>.

3. A division with a remainder of 0 is said to come out <u>even or exactly</u>.

5. A <u>prime</u> number is a whole number, greater than 1, that has only 1 and <u>itself</u> as factors.

7. An <u>even</u> whole number can be divided evenly by 2.

9. To prime <u>factor</u> a number means to write it as a product of only <u>prime</u> numbers.

11. In the exponential expression 6^4, 6 is called the <u>base</u>, and 4 is called the <u>exponent</u>.

CONCEPTS

13. $27 = 1 \cdot 27$ or $27 = 3 \cdot 9$

15. 2, 4, 22, 44, 11, 1 are the factors of 44.

17. The factors of 11 are 1 and 11.

19. The factors of 37 are 1 and 37.

21. Each of the numbers is prime.

23. Yes, if 4 is a factor of a whole number, then 4 will divide the number exactly.

25. $2 \cdot 3 \cdot 3 \cdot 5 = 90$

27. $11^2 \cdot 5 = 605$

29. No, you cannot change the order and obtain the same result, $3^2 = 9$ and $2^3 = 8$.

31. $30 = 2 \cdot 3 \cdot 5$; $242 = 2 \cdot 11 \cdot 11$
They have the prime factor 2 in common.

33. $20 = 2 \cdot 2 \cdot 5$;
$80 = 2 \cdot 2 \cdot 2 \cdot 2 \cdot 5$
They have prime factors 2 and 5 in common.

35.

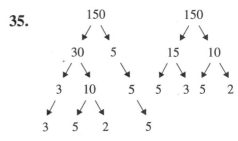

The first tree gives $3 \cdot 5 \cdot 2 \cdot 5$, the second tree gives $5 \cdot 3 \cdot 5 \cdot 2$. The order of the prime factors is different, but the prime factorization is the same.

37. $24 = 1 \cdot 24$ or $24 = 2 \cdot 12$
or $24 = 3 \cdot 8$ or $24 = 4 \cdot 6$

a. 6 and 4 are the factors whose sum is 10.

b. 2 and 12 are the factor whose difference is 10.

39. When using the division method to find the prime factorization of an even number, an obvious start would be to use 2.

NOTATION

41. $7^3 = 7 \cdot 7 \cdot 7$

43. $3^5 = 3 \cdot 3 \cdot 3 \cdot 3 \cdot 3$

45. $2 \cdot 2 \cdot 2 \cdot 2 \cdot 2 = 2^5$

47. $5 \cdot 5 \cdot 5 \cdot 5 = 5^4$

PRACTICE

49. factors of 10: 1, 2, 5, 10

51. factors of 40: 1, 2, 4, 5, 8, 10, 20, 40

53. factors of 18: 1, 2, 3, 6, 9, 18

55. factors of 44: 1, 2, 4, 11, 22, 44

57. factors of 77: 1, 7, 11, 77

59. $39 = 3 \cdot 13$

61. $30 = 2 \cdot 3 \cdot 5$

63. $162 = 2 \cdot 3^4$

65. $220 = 2^2 \cdot 5 \cdot 11$

67. $3^4 = 81$

69. $2^5 = 32$

71. $12^2 = 144$

73. $8^4 = 4,096$

75. $3^2 \cdot 2^3 = 72$

77. $2^3 \cdot 3^3 \cdot 4^2 = 3,456$

79. $234^3 = 12,812,904$

81. $23^2 \cdot 13^3 = 1,162,213$

APPLICATIONS

83. factors of 6: 1, 2, 3, 6;
$1 + 2 + 3 = 6$
factors of 28: 1, 2, 4, 7, 14, 28;
$1 + 2 + 4 + 7 + 14 = 28$

85. $5 = 5^1$; $25 = 5^2$; $125 = 5^3$;
$625 = 5^4$

REVIEW

91. $341 + 527 = 868$

93. $312 - 176 = 136$

95. $38 \cdot 42 = 1,596$

97. $615 \div 15 = 41$

99. $100 \cdot 44 + 100 \cdot 32$
$= 4,400 + 3,200$
$= 7,600$
He spent $7,600.

STUDY SET Section 1.5

VOCABULARY

1. Punctuation marks in mathematics are the grouping symbols.

3. The expression below a fraction bar is called the denominator.

CONCEPTS

5. To evaluate $5(2)^2 - 1$ requires 3 steps: square, multiply, subtract.

7. To evaluate $\dfrac{5+5(7)}{2+(8-4)}$ multiply first in the numerator and subtract first in the denominator.

9. In $2 \cdot 3^2$, square first:
$2 \cdot 3^2 = 2 \cdot 9 = 18$; in $(2 \cdot 3)^2$,
multiply first: $(2 \cdot 3)^2 = 6^2 = 36$

NOTATION

11. The symbols ( ) are called <u>parentheses</u>.

13. The symbols { } are called <u>braces</u>.

15. $28 - 5(2)^2 = 28 - 4(\,4\,)$
$\qquad = 28 - \overline{20}$
$\qquad = 8$

17. $[4 \cdot (2+7)] - 6 = (4 \cdot 9\,) - 6$
$\qquad = 36 \overline{} 6$
$\qquad = \overline{30}$

PRACTICE

19. $7 + 4 \cdot 5 = 7 + 20$
$\qquad = 27$

21. $10 + (8+5) = 10 + 13$
$\qquad = 23$

23. $20 - 10 + 5 = 10 - 5$
$\qquad = 5$

25. $15 - (8-3) = 15 - 5$
$\qquad = 10$

27. $7 \cdot 5 - 5 \cdot 6 = 35 - 30$
$\qquad = 5$

29. $4^2 + 3^2 = 16 + 9$
$\qquad = 25$

31. $2 \cdot 3^2 = 2 \cdot 9$
$\qquad = 18$

33. $3 + 2 \cdot 3^4 \cdot 5 = 3 + 2 \cdot 81 \cdot 5$
$\qquad = 3 + 162 \cdot 5$
$\qquad = 3 + 810$
$\qquad = 813$

35. $6 + 2(5+4) = 6 + 2(9)$
$\qquad = 6 + 18$
$\qquad = 24$

37. $3 + 5(6-4) = 3 + 5(2)$
$\qquad = 3 + 10$
$\qquad = 13$

39. $(7-4)^2 + 1 = 3^2 + 1$
$\qquad = 9 + 1$
$\qquad = 10$

41. $6^3 - (10-8) = 216 - (10-8)$
$\qquad = 216 - 2$
$\qquad = 214$

43. $50 - 2(4)^2 = 50 - 2(16)$
$\qquad = 50 - 32$.
$\qquad = 18$

45. $3^4 - (3)(2) = 81 - 6$
$\qquad = 75$

47. $39 - 5(6) + 9 - 1 = 39 - 30 + 9 - 1$
$\qquad = 9 + 9 - 1$
$\qquad = 18 - 1$
$\qquad = 17$

49. $(18-12)^3 - 5^2 = 6^3 - 5^2$
$\qquad = 216 - 25$
$\qquad = 191$

51. $2(10 - 3^2) + 1 = 2(10 - 9) + 1$
$$= 2(1) + 1$$
$$= 2 + 1$$
$$= 3$$

53. $6 + \dfrac{25}{5} + 6 \cdot 3 = 6 + 5 + 18$
$$= 11 + 18$$
$$= 29$$

55. $3\left(\dfrac{18}{3}\right) - 2(2) = 3(6) - 2(2)$
$$18 - 4$$
$$= 14$$

57. $(2 \cdot 6 - 4)^2 = (12 - 4)^2$
$$= 8^2$$
$$= 64$$

59. $\dfrac{10 + 5}{6 - 1} = \dfrac{15}{5}$
$$= 3$$

61. $\dfrac{5^2 + 17}{6 - 2^2} = \dfrac{25 + 17}{6 - 4}$
$$= \dfrac{42}{2}$$
$$= 21$$

63. $\dfrac{(3 + 5)^2 + 2}{2(8 - 5)} = \dfrac{8^2 + 2}{2(3)}$
$$= \dfrac{64 + 2}{6}$$
$$= \dfrac{66}{6}$$
$$= 11$$

65. $\dfrac{(5 - 3)^2 + 2}{4^2 - (8 + 2)} = \dfrac{2^2 + 2}{16 - 10}$
$$= \dfrac{4 + 2}{6}$$
$$= \dfrac{6}{6}$$
$$= 1$$

67. $2,985 - (1,800 + 689) = 2,985 - 2,489$
$$= 496$$

69. $3,245 - 25(16 - 12)^2 = 3,245 - 25(4)^2$
$$= 3,245 - 25(16)$$
$$= 3,245 - 400$$
$$= 2,845$$

APPLICATIONS

71. $3 \cdot 5 + 5 \cdot 12 = 15 + 60$
$$= 75$$
Pat paid $75.

73. Throw out the 2 and the 6.

$$\text{score} = \frac{5+4+3+4}{4}$$
$$= \frac{16}{4}$$
$$= 4$$

The diver's score is 4.

75. $\text{average} = \dfrac{75+80+83+80+77+72+86}{7}$
$$= \frac{553}{7}$$
$$= 79$$

The week's mean temperature was $79°$.

77. $\text{average} = \dfrac{56+75+82+63+79}{5}$
$$= \frac{355}{5}$$
$$= 71$$

Andrew's average score was 71.

REVIEW

83.
$$\begin{array}{r} 325 \\ + 349 \\ \hline 674 \end{array}$$

85.
$$\begin{array}{r} 5,628 \\ - 4,509 \\ \hline 1,119 \end{array}$$

87.
$$\begin{array}{r} 417 \\ \times \; 23 \\ \hline 9,591 \end{array}$$

89. $43\overline{)31,175}$ quotient 725

STUDY SET Section 1.6

VOCABULARY

1. An equation is a statement that two expressions are <u>equal</u>.

3. The answer to an equation is called a <u>solution</u> or a <u>root</u>.

5. <u>Equivalent</u> equations have exactly the same solutions.

CONCEPTS

7. If $x = y$ and c is any number, then $x + c =$ <u>$y + c$</u>.

9. In the equation $x + 6 = 10$, the addition of 6 is performed on the variable. We undo it by subtracting 6 from both sides.

NOTATION

11.
$$x + 8 = 24$$
$$x + 8 - 8 = 24 - 8$$
$$x = 16$$

PRACTICE

13. $x = 2$
yes, an equation

15. $7x < 8$
no, not an equation

17. $x + y = 0$
yes, an equation

19. $1 + 1 = 3$
yes, an equation

21. $1 + 2 = 3$
$3 = 3$
yes, a solution

23. $7 - 7 = 0$
$0 = 0$
yes, a solution

25. $8 - 5 = 5$
$3 \neq 5$
no, not a solution

27. $16 + 32 = 0$
$48 \neq 0$
no, not a solution

29. $7 + 7 = 7$
$14 \neq 7$
no, not a solution

31. $0 = 0$
yes, a solution

33. $\begin{aligned} x - 7 &= 3 \\ x + 7 - 7 &= 3 + 7 \\ x &= 10 \end{aligned}$

$\underline{\text{check}}$: $\begin{aligned} 10 - 7 &\overset{?}{=} 3 \\ 3 &= 3 \end{aligned}$

35. $\begin{aligned} a - 2 &= 5 \\ a + 2 - 2 &= 5 + 2 \\ a &= 7 \end{aligned}$

37. $\begin{aligned} 1 &= b - 2 \\ 1 + 2 &= b + 2 - 2 \\ 3 &= b \end{aligned}$

39. $\begin{aligned} x - 4 &= 0 \\ x + 4 - 4 &= 0 + 4 \\ x &= 4 \end{aligned}$

41. $\begin{aligned} y - 7 &= 6 \\ y + 7 - 7 &= 6 + 7 \\ y &= 13 \end{aligned}$

43. $\begin{aligned} 70 &= x - 5 \\ 70 + 5 &= x + 5 - 5 \\ 75 &= x \end{aligned}$

45. $\begin{aligned} 312 &= x - 428 \\ 312 + 428 &= x + 428 - 428 \\ 740 &= x \end{aligned}$

47. $\begin{aligned} x - 117 &= 222 \\ x + 117 - 117 &= 222 + 117 \\ x &= 339 \end{aligned}$

49. $\begin{aligned} x + 9 &= 12 \\ x + 9 - 9 &= 12 - 9 \\ x &= 3 \end{aligned}$

51. $\begin{aligned} y + 7 &= 12 \\ y + 7 - 7 &= 12 - 7 \\ y &= 5 \end{aligned}$

53. $\begin{aligned} t + 19 &= 28 \\ t + 19 - 19 &= 28 - 19 \\ t &= 9 \end{aligned}$

55. $\begin{aligned} 23 + x &= 33 \\ 23 - 23 + x &= 33 - 23 \\ x &= 10 \end{aligned}$

57. $\begin{aligned} 5 &= 4 + c \\ 5 - 4 &= 4 - 4 + c \\ 1 &= c \end{aligned}$

59. $\begin{aligned} 99 &= r + 43 \\ 99 - 43 &= r + 43 - 43 \\ 56 &= r \end{aligned}$

61.
$$512 = x + 428$$
$$512 - 428 = x + 428 - 428$$
$$84 = x$$

63.
$$x + 117 = 222$$
$$x + 117 - 117 = 222 - 117$$
$$x = 105$$

65.
$$3 + x = 7$$
$$3 - 3 + x = 7 - 3$$
$$x = 4$$

67.
$$y - 5 = 7$$
$$y + 5 - 5 = 7 + 5$$
$$y = 12$$

69.
$$4 + a = 12$$
$$4 - 4 + a = 12 - 4$$
$$a = 8$$

71.
$$x - 13 = 34$$
$$x + 13 - 13 = 34 + 13$$
$$x = 47$$

APPLICATIONS

73. **A**: The age of the jar; 1,700 years; 425 years

F: Let x = the age of the jar
Key phrase: <u>older than</u>
Translation: <u>add</u>
The age of the manuscript is 425 plus the age of the jar.
<u>1,700</u> = 425 + x

S:
$$1,700 = 425 + x$$
$$1,700 - 425 = 425 + x - 425$$
$$1,275 = x$$

S: The jar is 1,275 years old.

75. Let x = the number of invitations Mia sent.
$$x - 3 = 59$$
$$x + 3 - 3 = 59 + 3$$
$$x = 62$$
Mia sent 62 invitations.

77. Let x = the amount the Kim paid.
$$x - 12 = 68$$
$$x + 12 - 12 = 68 + 12$$
$$x = 80$$
Kim paid $80.

79. Let x = how much Holly paid.
$$48 = x + 9$$
$$48 - 9 = x + 9 - 9$$
$$39 = x$$
Holly paid $39.

81. Let x = how much she paid.
$$x + 29 = 219$$
$$x + 29 - 29 = 219 - 29$$
$$x = 190$$
She paid $190 to have her car fixed.

REVIEW

87. 325,780

89.
$$2 \cdot 3^2 \cdot 5 = 2 \cdot 9 \cdot 5$$
$$= 18 \cdot 5$$
$$= 90$$

91.
$$8 - 2(3) + 1^3 = 8 - 2(3) + 1$$
$$= 8 - 6 + 1$$
$$= 2 + 1$$
$$= 3$$

STUDY SET Section 1.7

VOCABULARY

1. The statement "If equal quantities are divided by the same nonzero quantity, the results will be equal quantities" expresses the <u>division</u> property of equality.

CONCEPTS

3. To undo a multiplication by 6, we <u>divide</u> both sides of the equation by <u>6</u>.

5. In the equation $4t = 40$, the variable is being multiplied by 4. To undo it, divide by 4.

7. **a.** Subtract 5 from both sides.

   b. Add 5 to both sides.

   c. Divide both sides by 5.

   d. Multiply both sides by 5.

9. If $x = y$ then $\dfrac{x}{z} = \dfrac{y}{z}$

NOTATION

11. $3x = 12$
$$\frac{3x}{3} = \frac{12}{3}$$
$$x = 4$$

PRACTICE

13. $3x = 3$
$$\frac{3x}{3} = \frac{3}{3}$$
$$x = 1$$

15. $32x = 192$
$$\frac{32x}{32} = \frac{192}{32}$$
$$x = 6$$

17. $17y = 51$
$$\frac{17y}{17} = \frac{51}{17}$$
$$y = 3$$

19. $34y = 204$
$$\frac{34y}{34} = \frac{204}{34}$$
$$y = 6$$

21. $$\frac{x}{7} = 2$$
$$7 \cdot \frac{x}{7} = 7 \cdot 2$$
$$x = 14$$

23. $$\frac{y}{14} = 3$$
$$14 \cdot \frac{y}{14} = 14 \cdot 3$$
$$y = 42$$

25. $$\frac{a}{15} = 15$$
$$15 \cdot \frac{a}{15} = 15 \cdot 15$$
$$a = 225$$

27. $$\frac{c}{13} = 13$$
$$13 \cdot \frac{c}{13} = 13 \cdot 13$$
$$c = 169$$

29. $9z = 9$
$$\frac{9z}{9} = \frac{9}{9}$$
$$z = 1$$

31. $7x = 21$
$$\frac{7x}{7} = \frac{21}{7}$$
$$x = 3$$

33. $43t = 86$
$$\frac{43t}{43} = \frac{86}{43}$$
$$t = 2$$

35. $21s = 21$
$$\frac{21s}{21} = \frac{21}{21}$$
$$s = 1$$

37.
$$\frac{d}{20} = 2$$
$$20 \cdot \frac{d}{20} = 20 \cdot 2$$
$$d = 40$$

39.
$$400 = \frac{t}{3}$$
$$3 \cdot 400 = 3 \cdot \frac{t}{3}$$
$$1,200 = t$$

APPLICATIONS

41. **A:** The original cost of the stock; $12,500

    F: Let x = the original cost of the stock

        Key phrase: <u>only worth $\frac{1}{3}$ of his original purchase price</u>

        Translation: <u>division</u>
$$\frac{\text{the original cost of the stock}}{3} = \text{the current value of the stock}$$

$$\frac{x}{3} = \underline{\$12{,}500}$$

    S:
$$\frac{x}{3} = \underline{12{,}500}$$
$$3 \cdot \frac{x}{3} = 3 \cdot 12{,}500$$
$$x = \underline{37{,}500}$$

    S: The original cost of the stock was $37,500.

43. Let x = her rate after taking the class.
$$r = 3 \cdot 130$$
$$r = 390$$
She can expect to read 390 wpm after taking the class.

45. Let x = each employee's share.
$$\frac{480{,}000}{12} = x$$
$$40{,}000 = x$$
Each employee's share would be $40,000.

47. Let x = the number of people expected to attend.

$$\frac{1}{4}x = 120$$
$$4 \cdot \frac{1}{4}x = 4 \cdot 120$$
$$x = 480$$

480 people were expected to attend.

REVIEW

51. $P = w + w + l + l$
$= 8 + 8 + 16 + 16$
$= 48$

The perimeter is 48 cm.

53. $120 = 2^3 \cdot 3 \cdot 5$

55. $3^2 \cdot 2^3 = 9 \cdot 8$
$ = 72$

57. average $= \dfrac{76 + 80 + 74 + 83 + 72}{5}$
$= \dfrac{385}{5}$
$= 77$

KEY CONCEPT Variables

1. Let x = the monthly cost to lease the van.

3. Let x = the width of the field.

5. Let x = distance traveled by motorist.

7. $a + b = b + a$

9. $\dfrac{b}{1} = b$

11. $n - 1 < n$

CHAPTER 1 REVIEW

1. a. 2, 5, 9

   b. 0, 2, 5, 9

3. a. 6 is in the ten thousands column.

   b. 7 is in the hundreds column.

5. a. 3,207

   b. 23,253,412

7. a. 2,507,300

   b. 2,510,000

   c. 2,507,350

   d. 2,500,000

9. a. $135 + 213 = 348$

   b. $4,447 + 7,478 = 11,925$

   c. $\begin{array}{r} 236 \\ + 282 \\ \hline 518 \end{array}$

   d. $\begin{array}{r} 5,345 \\ + 655 \\ \hline 6,000 \end{array}$

11. a. $8 - 5 = 3$

   b. $9 - (7 - 2) = 9 - 5$
   $ = 4$

   c. $235 - 218 = 17$

   d. $5,231 - 5,177 = 54$

   e. $\begin{array}{r} 343 \\ - 269 \\ \hline 74 \end{array}$

11. f.
$$\begin{array}{r} 7,800 \\ -\,5,725 \\ \hline 2,075 \end{array}$$

13. $931 + 271 - (37 + 380) = 931 + 271 - 417$
$$= 1,202 - 417$$
$$= 785$$

The final balance is $785.

15. a. $8 \cdot 7 = 56$

   b. $7 \cdot 8 = 56$

   c. $8 \cdot 0 = 0$

   d. $7 \cdot 1 = 7$

   e. $(5 \cdot 7) \cdot 6 = 35 \cdot 6$
$$= 210$$

   f. $5 \cdot (7 \cdot 6) = 5 \cdot 42$
$$= 210$$

17. $38 \cdot 9 = 342$
She earned $342.

19. Sarah: $12 \cdot 7 = 84$
Santiago: $15 \cdot 6 = 90$
Santiago earned the most money.

21. a. $357 \div 17 = 21$

   b. $1,443 \div 39 = 37$

   c. $21\overline{)405}$ 19R6

   d. $54\overline{)1,269}$ 23R27

23. There will be 25 candles left over.

25. a. 31 is prime

   b. 100 is composite

   c. 1 is neither

   d. 0 is neither

   e. 125 is composite

   f. 47 is prime

27. a. $42 = 2 \cdot 3 \cdot 7$

   b. $75 = 3 \cdot 5^2$

29. a. $5^3 = 125$

   b. $11^2 = 121$

   c. $2^3 \cdot 5^2 = 200$

   d. $2^2 \cdot 3^3 \cdot 5^2 = 2,700$

31. a. average $= \dfrac{80 + 74 + 66 + 88}{4}$
$$= \dfrac{308}{4}$$
$$= 77$$

   b. $\dfrac{73 + 77 + 81 + 69 + 90}{5} = \dfrac{390}{5}$
$$= 78$$

33. a. $y - 12 = 50$ The variable is y.

   b. $114 = 4 - t$ The variable is t.

35. a. $\begin{aligned} x + 9 &= 18 \\ x + 9 - 9 &= 18 = 9 \\ x &= 9 \end{aligned}$

   b. $\begin{aligned} b + 12 &= 26 \\ b + 12 - 12 &= 26 - 12 \\ b &= 14 \end{aligned}$

   c. $\begin{aligned} 175 &= p + 55 \\ 175 - 55 &= p + 55 - 55 \\ 120 &= p \end{aligned}$

   d. $\begin{aligned} 212 &= m + 207 \\ 212 - 207 &= m + 207 - 207 \\ 5 &= m \end{aligned}$

37. Let x = the number of patients originally.
$$\begin{aligned} x - 13 &= 172 \\ x + 13 - 13 &= 172 + 13 \\ x &= 185 \end{aligned}$$
He originally had 185 patients.

39. a. $\begin{aligned} \frac{x}{7} &= 3 \\ 7 \cdot \frac{x}{7} &= 7 \cdot 3 \\ x &= 21 \end{aligned}$

   b. $\begin{aligned} \frac{x}{3} &= 12 \\ 3 \cdot \frac{a}{3} &= 3 \cdot 12 \\ a &= 36 \end{aligned}$

   c. $\begin{aligned} 15 &= \frac{s}{21} \\ 21 \cdot 15 &= 21 \cdot \frac{s}{21} \\ 315 &= s \end{aligned}$

39. d. $\begin{aligned} 25 &= \frac{d}{17} \\ 17 \cdot 25 &= 17 \cdot \frac{d}{17} \\ 425 &= d \end{aligned}$

41. Let x = cost of the chain.
$$\begin{aligned} \frac{x}{4} &= 32 \\ 4 \cdot \frac{x}{4} &= 4 \cdot 32 \\ x &= 128 \end{aligned}$$
The chain cost $128.

CHAPTER 1 TEST

1. Whole numbers less than 5: 0, 1, 2, 3, 4

3. 7,507

5.

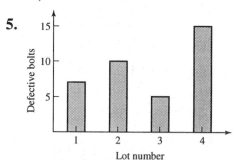

7. $15 \geq 10$

9. $327 + 435 = 762$

11. $\begin{aligned} 4&,521 \\ + 3&,579 \\ \hline 8&,100 \end{aligned}$

13. $\begin{aligned} P &= 327 + 757 + 327 + 757 \\ &= 2,168 \end{aligned}$
Perimeter is 2,168 in.

15. $\begin{aligned} 53 \\ \times\ 8 \\ \hline 424 \end{aligned}$

17.

$$63\overline{)4{,}536}$$ with quotient 72 shown above

$$
\begin{array}{r}
72 \\
63\overline{)4{,}536}
\end{array}
$$

19. $P = 23 + 17 + 23 + 17$
$= 80$
$A = 23 \cdot 17$
$= 391$
The perimeter is 80 cm; area is 391 cm^2.

21. $1{,}260 = 2^2 \cdot 3^2 \cdot 5 \cdot 7$

23. $9 + 4 \cdot 5 = 9 + 20$
$= 29$

25. $\dfrac{73 + 52 + 70 + 0 + 0}{5} = \dfrac{195}{5}$
$= 39$

27. $10 = x + 6$
$10 - 6 = x + 6 - 6$
$4 = x$

29. $5t = 55$
$\dfrac{5t}{5} = \dfrac{55}{5}$
$t = 11$

31. $500 + x = 700$
$500 - 500 + x = 700 - 500$
$x = 200$

33. Let $x =$ the number of parking spaces now.
$2x = 6{,}200$
$\dfrac{2x}{2} = \dfrac{6{,}200}{2}$
$x = 3{,}100$
The college has 3,100 parking spaces now.

VOCABULARY

1. Numbers can be represented by points equally spaced on a <u>number</u> <u>line</u>.

3. To <u>graph</u> a number means to locate it on a number line and highlight it with a <u>dot</u>.

5. The symbols > and < are called <u>inequality</u> symbols.

7. The <u>absolute</u> <u>value</u> of a number is the distance between it and zero on a number line.

9. The collection of all whole number and their opposites is called the <u>integers</u>.

CONCEPTS

11. **a.** $y \leq x$

   b. $x \geq y$

   c. $x \leq 0$ and $y \geq 0$

13. Yes, every number on the number line has an opposite.

15. $15 - 8$ contains a minus sign.

17. $15 > 12$

19. **a.** −225; $225 deposit

   b. −10; 10 seconds after liftoff

19. **c.** −3; 3 degrees above normal

   d. −12,000; a surplus of $12,000

   e. −2; 2 lengths ahead of the second place finisher

21. x is negative.

23. On a number line, − 4 is 3 units to the right of −7.

25. − 8 and 2 are 5 units away from -3.

27. −7 is closer to -3 on the number line.

29. $6 - 4$, $- 6$, $- (- 6)$ use the − symbol in three different ways.

PRACTICE

31. $|9| = 9$

33. $|- 8| = 8$

35. $|-14| = 14$

37. $-|20| = - 20$

39. $-|-6| = - 6$

41. $|203| = 203$

43. $- (- 4) = 4$

45. $- (- 12) = 12$

47.

49.

51. $-5 \leq 5$

53. $-12 \leq -6$

55. $-10 \geq -11$

57. $|-2| \geq 0$

59. $-1{,}255 \leq -(-1{,}254)$

61. $-|-3| \leq 4$

63. Let p = pounds collected this year.
$p > 391$

65. Let d = depth of the submarine.
$d < -500$

67. Let t = race time.
$t < 50$

69. Let a = altitude of the town.

71.

Time	Position of ball
1 sec	2
2 sec	3
3 sec	2
4 sec	0
5 sec	−3
6 sec	−7

73. The peaks are 2, 4, and 0. The valleys are −3, −5, and −2.

75. a. −1 (1 below par)

  b. −2 (2 below par)

  c. Most of the scores are below par.

77. a. The temperatures range from −10° to −20°.

77. b. It will be 10° colder in Chicago.

c. The coldest it should get in Seattle is 10°.

79. a. The time line uses a scale of 200 yr.

b. The origin represents the birth of Christ.

c. The positive numbers are A.D.

d. The negative numbers are B.C.

81.

REVIEW

87. 23,500

89.
$$2x = 34$$
$$\frac{2x}{2} = \frac{34}{2}$$
$$x = 17$$

91. associative property of multiplication

STUDY SET Section 2.2

VOCABULARY

1. When 0 is added to a number, the number remains the same. We call 0 the additive <u>identity</u>.

CONCEPTS

3. $-3 + 6 = 3$

5. $-5 + 3 = -2$

7. Yes, the sum of two positive integers is always positive.

9. The sum of a number and its additive inverse is 0.

11. The commutative property of addition an opposite.

13. To add two integers with unlike signs, <u>subtract</u> their absolute values, the <u>smaller</u> from the larger. Then attach to that result the sign of the number with the <u>larger</u> absolute value.

NOTATION

15. $-16 + (-2) + (-1) = \underline{-18} + (-1)$
$$= -19$$

17. $(-3 + 8) + (-3) = \underline{5} + -3$
$$= 2$$

PRACTICE

19. 11

21. 23

23. 0

25. −14

27. $6 + (+3) = 9$

29. $-5 + (-5) = -10$

31. $-6 + 7 = 1$

33. $-15 + 8 = -7$

35. $20 + (-40) = -20$

37. $30 + (-15) = 15$

39. $-1 + 9 = 8$

41. $-7 + 9 = 2$

43. $5 + (-15) = -10$

45. $24 + (-15) = 9$

47. $35 + (-27) = 8$

49. $24 + (-45) = -21$

51. $\begin{aligned} -2 + 6 + (-1) &= 4 + (-1) \\ &= 3 \end{aligned}$

53. $\begin{aligned} -9 + 1 + (-2) &= -8 + (-2) \\ &= -10 \end{aligned}$

55. $\begin{aligned} 6 + (-4) + (-13) + 7 &= 2 + (-13) + 7 \\ &= -11 + 7 \\ &= -4 \end{aligned}$

57. $\begin{aligned} 9 + (-3) + 5 + (-4) &= 6 + 5 + (-4) \\ &= 11 + (-4) \\ &= 7 \end{aligned}$

59. $\begin{aligned} -6 + (-7) + (-8) &= -13 + (-8) \\ &= -21 \end{aligned}$

61. $-7 + 0 = -7$

63. $9 + 0 = 9$

65. $-4 + 4 = 0$

67. $2 + (-2) = 0$

69. 5 must be added to -5 to obtain 0.

71. $\begin{aligned} 2 + (-10 + 8) &= 2 + (-2) \\ &= 0 \end{aligned}$

73. $\begin{aligned} (-4 + 8) + (-11 + 4) &= 4 + (-7) \\ &= -3 \end{aligned}$

75. $\begin{aligned} [-3 + (-4)] + (-5 + 2) &= -7 + (-3) \\ &= -10 \end{aligned}$

77. $\begin{aligned} [6 + (-4)] + [8 + (-11)] &= 2 + (-3) \\ &= -1 \end{aligned}$

79. $-2 + [-8 + (-7)] = -2 + (-15)$
$\qquad\qquad\qquad\quad = -17$

APPLICATIONS

81. $1G + 2G = 3G;$
$1G + (-4G) = -3G$

83. $-900 + 450 + 380 = -450 + 380$
$\qquad\qquad\qquad\qquad = -70$
No, there is a $70 shortfall each month.

85. a. -85 represents 85 ft below sea level.

b. 20 ft was reeled in after each reading.

85. c. The greatest temperature change was between -45 and -65 ft.

87. a. $-4 + 3 = -1$

b. $-4 + 4 = 0$

89. $-4 + 11 = 7$
The river will be 7 ft over flood stage.

91. $789 + (-9,135) = -8,346$

93. $-675 + (-456) + 99 = -1,032$

95. $10 + (-5) + 15 + (-10) = 5 + 15 + (-10)$
$\qquad\qquad\qquad\qquad\quad = 20 + (-10)$
$\qquad\qquad\qquad\qquad\quad = 10$
The studio had a profit of $10 million.

REVIEW

101. $A = 5 \cdot 3$
$\quad\; = 15$
The area is 15 ft^2.

103. $\qquad x - 7 = 20$
$\quad x - 7 + 7 = 20 + 7$
$\qquad\qquad x = 27$

105. $125 = 5^3$

STUDY SET Section 2.3
VOCABULARY

1. The answer to a subtraction problem is called the <u>difference</u>.

3. Two numbers represented by points on a number line that are the same distance away from the origin, but on opposite sides of it, are called <u>opposites</u>.

CONCEPTS

5. <u>Subtraction</u> is the same as adding the opposite of the number to be subtracted.

7. Subtracting –6 is the same as adding <u>6.</u>

9. Every subtraction problem can be written as an equivalent <u>addition</u> problem.

11. After using parentheses as grouping symbols, if another set of grouping symbols is needed, we use <u>brackets.</u>

13. Negative eight minus negative four is written $-8 - (-4)$.

15. The distance between the points -4 and 3 is $3 - (-4) = 7$.

NOTATION

17.
$$1 - 3 - (-2) = 1 + \underline{(-3)} + 2$$
$$= -2 + \underline{2}$$
$$= 0$$

19.
$$(-8 - 2) - (-6) = [-8 + \underline{(-2)}] - (-6)$$
$$= -10 - (-6)$$
$$= -10 + \underline{6}$$
$$= -4$$

PRACTICE

21.
$$8 - (-1) = 8 + 1$$
$$= 9$$

23.
$$-4 - 9 = -4 + (-9)$$
$$= -13$$

25.
$$-5 - 5 = -5 + (-5)$$
$$= -10$$

27.
$$-5 - (-4) = -5 + 4$$
$$= -1$$

29.
$$-1 - (-1) = -1 + 1$$
$$= 0$$

31.
$$-2 - (-10) = -2 + 10$$
$$= 8$$

33.
$$0 - (-5) = 0 + 5$$
$$= 5$$

35.
$$0 - 4 = 0 + (-4)$$
$$= -4$$

37.
$$-2 - 2 = -2 + (-2)$$
$$= -4$$

39.
$$-10 - 10 = -10 + (-10)$$
$$= -20$$

41.
$$9 - 9 = 9 + (-9)$$
$$= 0$$

43.
$$-3 - (-3) = -3 + 3$$
$$= 0$$

45. $-4-(-4)-15 = -4+4+(-15)$
$\qquad\qquad\qquad = 0+(-15)$
$\qquad\qquad\qquad = -15$

47. $-3-3-3 = -3+(-3)+(-3)$
$\qquad\qquad\quad = -6+(-3)$
$\qquad\qquad\quad = -9$

49. $5-9-(-7) = 5+(-9)+7$
$\qquad\qquad\quad = -4+7$
$\qquad\qquad\quad = 3$

51. $10-9-(-8) = 10+(-9)+8$
$\qquad\qquad\qquad = 1+8$
$\qquad\qquad\qquad = 9$

53. $-1-(-3)-4 = -1+3+(-4)$
$\qquad\qquad\qquad = 2+(-4)$
$\qquad\qquad\qquad = -2$

55. $-5-8-(-3) = -5+(-8)+3$
$\qquad\qquad\qquad = -13+3$
$\qquad\qquad\qquad = -10$

57. $(-6-5)-3 = [-6+(-5)]-3$
$\qquad\qquad\quad = -11-3$
$\qquad\qquad\quad = -11+(-3)$
$\qquad\qquad\quad = -14$

59. $(6-4)-(1-2) = [6+(-4)]-[1+(-2)]$
$\qquad\qquad\qquad\quad = 2-(-1)$
$\qquad\qquad\qquad\quad = 2+1$
$\qquad\qquad\qquad\quad = 3$

61. $-9-(6-7) = -9-[6+(-7)]$
$\qquad\qquad\quad = -9-(-1)$
$\qquad\qquad\quad = -9+1$
$\qquad\qquad\quad = -8$

63. $-8-[4-(-6)] = -8-(4+6)$
$\qquad\qquad\qquad\quad = -8-10$
$\qquad\qquad\qquad\quad = -8+(-10)$
$\qquad\qquad\qquad\quad = -18$

65. $[-4+(-8)]-(-6) = -12-(-6)$
$\qquad\qquad\qquad\qquad = -12+6$
$\qquad\qquad\qquad\qquad = -6$

67. $7-(-3) = 7+3$
$\qquad\qquad = 10$

69. $-10-(-6) = -10+6$
$\qquad\qquad\quad = -4$

APPLICATIONS

71. $-50 - 70 = -50 + (-70)$
$= -120$
The diver's final depth is -120 ft.

73. $-7 - (-23) = -7 + 23$
$= 16$
The school's reading score improved by 16 points.

75. $5 - 7 - 6 = [5 + (-7)] - 6$
$= -2 - 6$
$= -2 + (-6)$
$= -8$
The ammeter will register -8.

77. $-1,290 - (-283) = -1,290 + 283$
$= -1,007$
The Dead Sea is 1,007 ft lower than Death Valley.

79. $-1 - 6 - 5 + 8 = [-1 + (-6)] - 5 + 8$
$= -7 - 5 + 8$
$= -7 + (-5) + 8$
$= -12 + 8$
$= -4$
The net gain is -4 yd.

81. a.

b. $25 - (-12) = 25 + 12$
$= 37$
The total length is 37 ft.

83. $-1,557 - 890 = -2,447$

85. $20,007 - (-496) = 20,503$

87. $-162 - (-789) - 2,303 = -1,676$

89. $1,303 - 676 - 121 - 121 - 750 = -244$
No, he will be $244 overdrawn.

REVIEW

95. $5x = 15$

$$\frac{5x}{5} = \frac{15}{5}$$

$$x = 3$$

97. The factors of 20: 1, 2, 4, 5, 10, 20

99. $12(13) = 156$
It takes 156 oranges.

101. $4{,}502 = 4$ thousands + 5 hundreds + 2 ones

STUDY SET Section 2.4

VOCABULARY

1. In the multiplication –5(–4), the numbers –5 and –4 are called factors. The answer, 20, is called the product.

3. The numbers . . . –4, –3, –2, –1, 0, 1, 2, 3, 4, . . . are called integers.

5. In the expression -3^5, 3 is the base and 5 is the exponent.

CONCEPTS

7. The product of two integers with unlike signs is negative.

9. The commutative property of multiplication implies that –2(–3) = –3(–2).

11. –1(9) = –9. The result is the opposite of the number.

13. The possible combinations are positive•positive, positive•negative, negative•positive, negative•negative.

15. a. $(-5)^{13}$ would be negative.

  b. $(-3)^{20}$ would be positive.

17. a. $|-3| = 3$ **b.** $|12| = 12$

  c. $|-5| = 5$ **d.** $|9| = 9$

  e. $|10| = 10$ **f.** $|-25| = 25$

19.

Problem	Number of negative factors	Answer
–2(–2)	2	4
–2(–2)(–2)(–2)	4	16
–2(–2)(–2)(–2)(–2)(–2)	6	64

NOTATION

21. $-3(-2)(-4) = \underline{6}(-4)$
$\qquad\qquad = \overline{-}24$

PRACTICE

23. $-9(-6) = 54$

25. $-3 \cdot 5 = -15$

27. $12(-3) = -36$

29. $(-8)(-7) = 56$

31. $(-2)10 = -20$

33. $-40(3) = -120$

35. $-8(0) = 0$

37. $-1(-6) = 6$

39. $-7(-1) = 7$

41. $1(-23) = -23$

43. $-6(-4)(-2) = 24(-2)$
$\qquad\qquad\quad = -48$

45. $5(-2)(-4) = -10(-4)$
$\qquad\qquad\quad = 40$

47. $2(3)(-5) = 6(-5)$
$\qquad\qquad = -30$

49. $6(-5)(2) = -30(2)$
$\qquad\qquad = -60$

51. $(-1)(-1)(-1) = 1(-1)$
$\qquad\qquad\qquad = -1$

53. $-2(-3)(3)(-1) = 6(3)(-1)$
$\qquad\qquad\qquad = 18(-1)$
$\qquad\qquad\qquad = -18$

55. $3(-4)(0) = -12(0)$
$\qquad\qquad = 0$

57. $-2(0)(-10) = 0(-10)$
$\qquad\qquad\quad = 0$

59. $-6(-10) = 60$

61. $(-4)^2 = 16$

63. $(-5)^3 = -125$

65. $(-2)^3 = -8$

67. $(-9)^2 = 81$

69. $(-1)^5 = -1$

71. $(-1)^8 = 1$

73. $(-7)^2 = (-7)(-7)$
$\qquad\quad = 49$

75. $-12^2 = -(12 \cdot 12)$
$\qquad\quad = -144$
$\quad(-12)^2 = (-12)(-12)$
$\qquad\qquad = 144$

APPLICATIONS

77. a. plan #1: $3(10) = 30$ lb
plan #2: $2(14) = 28$ lb

b. The patient should expect to lose the most weight with plan #1 but the workout time is double that of plan #2.

79. a. The high is 2; the low is –3.

b.

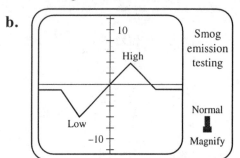

81. $5(-4) = -20$; $-20°$ total change

83. $10(-2) = -20$; -20 ft

85. $-76(787) = -59,812$

87. $(-81)^4 = 43,046,721$

89. $(-32)(-12)(-67) = -25,728$

91. $(-25)^4 = 390,625$

93. $(-3)(14,505) = -\$43,515$

REVIEW

99. $3^2 \cdot 5 = 9 \cdot 5$
$= 45$

101. $12,300 - 10,200 = 2,100$
The increase was 2,100.

103. $P = 4(6)$
$= 24$
The perimeter is 24 yd.

105. The first ten prime numbers: 2, 3, 5, 7, 11, 13, 17, 19, 23, 29

STUDY SET Section 2.5

VOCABULARY

1. In $\dfrac{-27}{3} = -9$, the number –9 is called the _quotient_, and the number 3 is the _divisor_.

3. The _absolute value_ of a number is the distance between it and 0 on the number line.

CONCEPTS

5. $5(-5) = -25$

7. $0(?) = -6$

9. $\dfrac{-20}{5} = -4$ or $\dfrac{-20}{-4} = 5$

11. The quotient of two negative integers is _positive._

13. a. always true

b. sometimes true

c. always true

PRACTICE

15. $\dfrac{-14}{2} = -7$

17. $\dfrac{-8}{-4} = 2$

19. $\dfrac{-25}{-5} = 5$

21. $\dfrac{-45}{-15} = 3$

23. $\dfrac{40}{-2} = -20$

25. $\dfrac{50}{-25} = -2$

27. $\dfrac{0}{-16} = 0$

29. $\dfrac{-6}{0}$ is undefined

31. $\dfrac{-5}{1} = -5$

33. $-5 \div (-5) = 1$

35. $\dfrac{-9}{9} = -1$

37. $\dfrac{-10}{-1} = 10$

39. $\dfrac{-100}{25} = -4$

41. $\dfrac{75}{-25} = -3$

43. $\dfrac{-500}{-100} = 5$

45. $\dfrac{-200}{50} = -4$

47. $\dfrac{-45}{9} = -5$

49. $\dfrac{8}{-2} = -4$

51. $\dfrac{-20}{5} = -4$
There was a 4° temperature drop per hour.

53. $\dfrac{-3,000}{3} = -1,000$
Each of the dives will be 1,000 ft.

55. $\dfrac{-12}{2} = -6$
They expect to finish 6 games behind.

57. $\dfrac{-300}{20} = -15$
She can mark down each pair $15.

59. $\dfrac{-13,550}{25} = -542$

61. $\dfrac{272}{-17} = -16$

63. $\dfrac{-9,135,000}{5,250} = -1,740$
Each employee will experience a $1,740 pay cut.

69. $3\left(\dfrac{18}{3}\right) - 2(2) = 3(6) - 2(2)$

$\qquad = 18 - 4$

$\qquad = 14$

71. $210 = 2 \cdot 3 \cdot 5 \cdot 7$

73. $\qquad 99 = r + 43$

$\qquad 99 - 43 = r + 43 - 43$

$\qquad 56 = r$

75. $3^4 = 81$

STUDY SET Section 2.6
VOCABULARY

1. When asked to evaluate expressions containing more than one operation, we should apply the rules for <u>order</u> of <u>operations</u>.

3. Absolute value symbols, parentheses, and brackets are types of <u>grouping</u> symbols.

CONCEPTS

5. Three operations must be performed: power, multiplication, subtraction.

7. In the numerator, do multiplication first; in the denominator, do subtraction first.

9. In -3^2, the base is 3; in $(-3)^2$, the base is –3.

NOTATION

11. $-8 - 5(-2)^2 = -8 - 5(4)$

$\qquad = -8 - 20$

$\qquad = -8 + (-20)$

$\qquad = -28$

13. $[-4(2 + 7)] - 6 = [-4(9)] - 6$

$\qquad = -36 - 6$

$\qquad = -42$

PRACTICE

15. $(-3)^2 - 4^2 = 9 - 16$

$\qquad = 9 + (-16)$

$\qquad = -7$

17. $3^2 - 4(-2)(-1) = 9 - 4(-2)(-1)$

$\qquad = 9 - 8$

$\qquad = 1$

19. $(2 - 5)(5 + 2) = -3(7)$

$\qquad = -21$

21. $-10 - 2^2 = -10 - 4$

$\qquad = -14$

23. $\dfrac{-6 - 8}{2} = \dfrac{-14}{2}$

$\qquad = -7$

25. $\dfrac{-5 - 5}{2} = \dfrac{-10}{2}$

$\qquad = -5$

27. $-12 \div (-2) \cdot 2 = 6 \cdot 2$

$\qquad = 12$

29. $-16 - 4 \div (-2) = -16 - (-2)$

$\qquad = -16 + 2$

$\qquad = -14$

31. $|-5(-6)| = |30|$
$= 30$

33. $|-4-(-6)| = |-4+6|$
$= |2|$
$= 2$

35. $(7-5)^2 - (1-4)^2 = (2)^2 - (-3)^2$
$= 4 - 9$
$= -5$

37. $-1(2^2 - 2 + 1^2) = -1(4 - 2 + 1)$
$= -1(2 + 1)$
$= -1(3)$
$= -3$

39. $-50 - 2(-3)^3 = -50 - 2(-27)$
$= -50 - (-54)$
$= -50 + 54$
$= 4$

41. $-6^2 + 6^2 = -36 + 36$
$= 0$

43. $3\left(\dfrac{-18}{3}\right) - 2(-2) = 3(-6) - 2(-2)$
$= -18 - (-4)$
$= -18 + 4$
$= -14$

45. $6 + \dfrac{25}{-5} + 6 \cdot 3 = 6 + (-5) + 18$
$= 1 + 18$
$= 19$

47. $\dfrac{1 - 3^2}{-2} = \dfrac{1 - 9}{-2}$
$= \dfrac{-8}{-2}$
$= 4$

49. $\dfrac{-4(-5) - 2}{-6} = \dfrac{20 - 2}{-6}$
$= \dfrac{18}{-6}$
$= -3$

51. $-3\left(\dfrac{32}{-4}\right) - (-1)^5 = -3(-8) - (-1)^5$
$= -3(-8) - (-1)$
$= 24 - (-1)$
$= 24 + 1$
$= 25$

53. $6(2^3)(-1) = 6(8)(-1)$
$= -48$

55. $2 + 3[5 - (1 - 10)] = 2 + 3[5 - (-9)]$
$= 2 + 3(14)$
$= 2 + 42$
$= 44$

57. $-7(2 - 3 \cdot 5) = -7(2 - 15)$
$= -7(-13)$
$= 91$

59. $-[6 - (1 - 4)] = -[6 - (-3)]$
$$= -(6 + 3)$$
$$= -(9)$$
$$= -9$$

61. $15 + (-3 \bullet 4 - 8) = 15 + (-12 - 8)$
$$= 15 + (-20)$$
$$= -5$$

63. $|-3 \bullet 4 + (-5)| = |-12 + -5|$
$$= |-17|$$
$$= 17$$

65. $\left|(-5)^2 - 2 \bullet 7\right| = |25 - 14|$
$$= |11|$$
$$= 11$$

67. $-379 + (-103) + 287 = -400 + (-100) + 300$
$$= -500 + 300$$
$$= -200$$

69. $-39 \bullet 8 = -40 \bullet 8$
$$= -320$$

71. $-3,887 + (-5,106) = -4,000 + (-5,000)$
$$= -9,000$$

73. $\dfrac{6,267}{-5} = \dfrac{6,000}{-5}$
$$= -1,200$$

APPLICATIONS

75. $12(3) + 3(-4) + 5(-1) = 36 + (-12) + (-5)$
$$= 36 + (-17)$$
$$= 19$$

77. $\dfrac{16 + 10 + 4 + (-2) + 0 + (-4)}{6} = \dfrac{30 + (-6)}{6}$
$$= \dfrac{24}{6}$$
$$= 4$$

He averaged 4 yards gained per carry.

79. $68 + 91 + (-47) + (-22) + (-34) = 70 + 90 + (-50) + (-20) + (-30)$
$$= 160 + (-100)$$
$$= 60$$

The value of a barrel of crude oil had a net gain of 60 cents that week.

81. $-2(-34)^2 - (-605) = -1,707$

83. $-60 - \dfrac{1{,}620}{-36} = -15$

REVIEW

89. $8 = 2x$

$\dfrac{8}{2} = \dfrac{2x}{2}$

$4 = x$

91. To find the perimeter of a rectangle, add the lengths of all its sides.

93. A sum is the result an addition; a difference is the result of a subtraction.

95. The factors of 36: 1, 2, 3, 4, 6, 9, 12, 18, 36.

STUDY SET Section 2.7

VOCABULARY

1. To <u>solve</u> an equation, we isolate the variable on one side of the equals sign.

CONCEPTS

3. If we multiply x by 3 and then divide that product by 3, the result is x.

5. $-10x$ means $-10 \cdot x$.

7. a. $x + 3 - 3 = 10 - \underline{3}$

b. $x + 3 + (-3) = 10 + \underline{(-3)}$

9. a. multiplication by -2

b. addition of -6

c. multiplication by -4; subtraction of 8

9. d. multiplication by -5; addition of -3

11. When solving the equation $-4 + t = -8 - 2$, it is best to <u>simplify</u> the right-hand side of the equation first before undoing any operations performed on the variable.

13. When solving an equation, we isolate the variable by undoing the operations performed on it in the <u>opposite</u> order.

15. a. Undo the subtraction of 3 first.

b. Undo the addition of -6 first.

NOTATION

17. $y + (-7) = -16 + 3$

$y + (-7) = \underline{-13}$

$y + (-7) + 7 = -13 + \underline{7}$

$y = -6$

19. $-13 = -4y - 1$

$-13 + \underline{1} = -4y - 1 + \underline{1}$

$\underline{-12} = -4y$

$\dfrac{-12}{-4} = \dfrac{-4y}{-4}$

$\underline{3} = y$

$y = 3$

PRACTICE

21. $-3x - 4 = 2$

$-3(-2) - 4 \overset{?}{=} 2$

$6 - 4 \overset{?}{=} 2$

$2 = 2$

Yes, -2 is a solution.

23. $-x+8=-4$

$-4+8\overset{?}{=}-4$

$4\neq-4$

No, 4 is not a solution.

25. $x+6=-12$

$x+6-6=-12-6$

$x=-18$

Check: $x+6=-12$

$-18+6\overset{?}{=}-12$

$-12=-12$

27. $-6+m=-20$

$-6+m+6=-20+6$

$m=-14$

29. $-5+3=-7+f$

$-2=-7+f$

$-2+7=-7+f+7$

$5=f$

$f=5$

31. $h-8=-9$

$h-8+8=-9+8$

$h=-1$

33. $0=y+9$

$0-9=y+9-9$

$-9=y$

$y=-9$

35. $r-(-7)=-1-6$

$r+7=-7$

$r+7-7=-7-7$

$r=-14$

37. $t-4=-8-(-2)$

$t-4=-6$

$t-4+4=-6+4$

$t=-2$

39. $x-5=-5$

$x-5+5=-5+5$

$x=0$

41. $-2s=16$

$\dfrac{-2s}{-2}=\dfrac{16}{-2}$

$s=-8$

Check: $-2s=16$

$-2(-8)\overset{?}{=}16$

$16=16$

43. $-5t=-25$

$\dfrac{-5t}{-5}=\dfrac{-25}{-5}$

$t=5$

45. $-2+(-4)=-3n$

$-6=-3n$

$\dfrac{-6}{-3}=\dfrac{-3n}{-3}$

$2=n$

$n=2$

47. $-9h=-8+17$

$-9h=9$

$\dfrac{-9h}{-9}=\dfrac{9}{-9}$

$h=-1$

49. $\dfrac{t}{-3}=-2$

$-3\left(\dfrac{t}{-3}\right)=-3(-2)$

$t=6$

Check: $\dfrac{t}{-3}=-2$

$\dfrac{6}{-3}\overset{?}{=}-2$

$-2=-2$

51.
$$0 = \frac{y}{8}$$
$$8(0) = 8\left(\frac{y}{8}\right)$$
$$0 = y$$
$$y = 0$$

53.
$$\frac{x}{-2} = -6 + 3$$
$$\frac{x}{-2} = -3$$
$$-2\left(\frac{x}{-2}\right) = -2(-3)$$
$$x = 6$$

55.
$$\frac{x}{4} = -5 - 8$$
$$\frac{x}{4} = -13$$
$$4\left(\frac{x}{4}\right) = 4(-13)$$
$$x = -52$$

57.
$$-2y + 8 = -6$$
$$-2y + 8 - 8 = -6 - 8$$
$$-2y = -14$$
$$\frac{-2y}{-2} = \frac{-14}{-2}$$
$$y = 7$$
Check: $-2y + 8 = -6$
$$-2(7) + 8 \overset{?}{=} -6$$
$$-14 + 8 \overset{?}{=} -6$$
$$-6 = -6$$

59.
$$-21 = -4h - 5$$
$$-21 + 5 = -4h - 5 + 5$$
$$-16 = -4h$$
$$\frac{-16}{-4} = \frac{-4h}{-4}$$
$$4 = h$$
$$h = 4$$

61.
$$-3v + 1 = 16$$
$$-3v + 1 - 1 = 16 - 1$$
$$-3v = 15$$
$$\frac{-3v}{-3} = \frac{15}{-3}$$
$$v = -5$$

63.
$$8 = -3x + 2$$
$$8 - 2 = -3x + 2 - 2$$
$$6 = -3x$$
$$\frac{6}{-3} = \frac{-3x}{-3}$$
$$-2 = x$$
$$x = -2$$

65.
$$-35 = 5 - 4x$$
$$-35 - 5 = 5 - 4x - 5$$
$$-40 = -4x$$
$$\frac{-40}{-4} = \frac{-4x}{-4}$$
$$10 = x$$
$$x = 10$$

67.
$$4 - 5x = 34$$
$$4 - 5x - 4 = 34 - 4$$
$$-5x = 30$$
$$\frac{-5x}{-5} = \frac{30}{-5}$$
$$x = -6$$

69.
$$-5-6-5x = 4$$
$$-11-5x = 4$$
$$-11-5x+11 = 4+11$$
$$-5x = 15$$
$$\frac{-5x}{-5} = \frac{15}{-5}$$
$$x = -3$$

71.
$$4-6x = -5-9$$
$$4-6x = -14$$
$$4-6x-4 = -14-4$$
$$-6x = -18$$
$$\frac{-6x}{-6} = \frac{-18}{-6}$$
$$x = 3$$

73.
$$\frac{h}{-6}+4 = 5$$
$$\frac{h}{-6}+4-4 = 5-4$$
$$\frac{h}{-6} = 1$$
$$-6\left(\frac{h}{-6}\right) = -6(1)$$
$$h = -6$$

75.
$$-3+(-5) = \frac{t}{-6}+1$$
$$-8 = \frac{t}{-6}+1$$
$$-8-1 = \frac{t}{-6}+1-1$$
$$-9 = \frac{t}{-6}$$
$$-6(-9) = -6\left(\frac{t}{-6}\right)$$
$$54 = t$$
$$t = 54$$

77.
$$0 = 6+\frac{c}{-5}$$
$$0-6 = 6+\frac{c}{-5}-6$$
$$-6 = \frac{c}{-5}$$
$$-5(-6) = -5\left(\frac{c}{-5}\right)$$
$$30 = c$$
$$c = 30$$

79.
$$-1 = -8+\frac{h}{-2}$$
$$-1+8 = -8+\frac{h}{-2}+8$$
$$7 = \frac{h}{-2}$$
$$-2(7) = -2\left(\frac{h}{-2}\right)$$
$$-14 = h$$
$$h = -14$$

81.
$$-x = 8$$
$$-1x = 8$$
$$(-1)(-1x) = (-1)8$$
$$x = -8$$
Check:
$$-x = 8$$
$$-1x = 8$$
$$(-1)(-8) \stackrel{?}{=} 8$$
$$8 = 8$$

83.
$$-15 = -k$$
$$-15 = -1k$$
$$(-1)(-15) = (-1)(-1k)$$
$$15 = k$$
$$k = 15$$

APPLICATIONS

85. **A:** <u>The number of feet cage was raised</u>

F: Let x = <u>the number of feet the cage was raised</u>

Key word: <u>raised</u>

Translation: <u>add</u>

$$\underline{-120} + \underline{\ x\ } = -75$$

S: $$-120 + x = \underline{-75}$$
$$-120 + x + \underline{120} = \overline{-75} + \underline{120}$$
$$x = 45$$

S: The shark cage was raised 45 feet.

C: $-120 + \underline{\ 45\ } = \underline{\ -75}$

87. Let x = the number of feet to be dredged out.
$$-47 - x = -65$$
$$-47 - x + 47 = -65 + 47$$
$$-x = -18$$
$$-1(-1x) = -1(-18)$$
$$x = 18$$
18 ft must be dredged out.

89. Let x = the number of rushing yards in the second half.
$$43 + x = -8$$
$$43 + x - 43 = -8 - 43$$
$$x = -51$$
Their rushing total in the second half was –51 yd.

91. Let x = the increase in market share over the 5-year span.
$$-43 + x = -9$$
$$-43 + x + 43 = -9 + 43$$
$$x = 34$$
The company picked up 34 points over the 5-year span.

93. Let x = the amount the price dropped each month.
$$x = \frac{-60}{12}$$
$$x = -5$$
The price dropped $5 each month.

95. Let x = how many points he gained over the 6-month period.
$$-31 + x = -2$$
$$-31 + x + 31 = -2 + 31$$
$$x = 29$$
He gained 29 points over the 6 months.

97. Let x = the amount of profit in the second year.
$$-11,560 + x = 32,090$$
$$-11,560 + x + 11,560 = 32,090 + 11,560$$
$$x = 43,650$$
They had a profit of $43,650 in the second year.

REVIEW

103. $5^6 = 5 \cdot 5 \cdot 5 \cdot 5 \cdot 5 \cdot 5$

105.
$$7 + 3y = 43$$
$$7 + 3y - 7 = 43 - 7$$
$$3y = 36$$
$$\frac{3y}{3} = \frac{36}{3}$$
$$y = 12$$

107. $16 \div 8 = \dfrac{16}{8}$

109.
$$A = l \cdot w$$
$$= 5(2)$$
$$= 10$$

The area is 10 in.^{2}

CHAPTER 2 REVIEW

1. a.

b.

3. a. $x < 75$ mph

   b. $x > \$14$ million

5. a. $|-4| = 4$

   b. $|0| = 0$

   c. $|-43| = 43$

   d. $-|12| = -12$

7. a. $-(-12) = 12$

   b. the opposite of 8

   c. the opposite of -8 is -8.

   d. $-0 = 0$ is 8.

9. a. $-6 + (-4) = -10$

   b. $-2 + (-3) = -5$

   c. $-23 + (-60) = -83$

   d. $-1 + (-4) + (-3) = -8$

11. a. $-4 + 0 = -4$

   b. $0 + (-20) = -20$

   c. $-8 + 8 = 0$

   d. $3 + (-3) = 0$

13. $-100 + 35 = -65$
The water level was 65 ft below normal after the rain.

15. Subtracting a number is the same as <u>adding</u> the <u>opposite</u> of that number.

17. $-150 - 75 = -225$
The depth of the second discovery was -225 ft.

19. Alaska:
$$100 - (-80) = 100 + 80$$
$$= 180$$
The difference is $180°$.

Virginia:
$$110 - (-30) = 110 + 30$$
$$= 140$$
The difference is $140°$.

21. a. $(-6)(-2)(-3) = -36$
    b. $4(-3)3 = -36$
    c. $0(-7) = 0$
    d. $(-1)(-1)(-1)(-1) = 1$

23. a. $(-5)^2 = 25$
    b. $(-2)^5 = -32$
    c. $(-8)^2 = 64$
    d. $(-4)^3 = -64$

25. -2^2 has a base of 2
$$-2^2 = -(2 \cdot 2)$$
$$= -4$$
$(-2)^2$ has a base of -2
$$(-2)^2 = (-2)(-2)$$
$$= 4$$

27. a. $\dfrac{-14}{7} = -2$

    b. $\dfrac{25}{-5} = -5$

    c. $\dfrac{-64}{8} = -8$

    d. $\dfrac{-20}{-2} = 10$

29. $\dfrac{-12}{6} = -2$
Production time dropped 2 minutes each month.

31. a. $-4\left(\dfrac{15}{-3}\right) - 2^3 = -4\left(\dfrac{15}{-3}\right) - 8$
$$= -4(-5) - 8$$
$$= 20 - 8$$
$$= 12$$

31. b. $-20 + 2(12 - 5 \cdot 2) = -20 + 2(12 - 10)$
$$= -20 + 2(2)$$
$$= -20 + 4$$
$$= -16$$

c. $20 + 2[12 - (-7 + 5)] = 20 + 2[12 - (-2)]$
$$= 20 + 2(12 + 2)$$
$$= 20 + 2(14)$$
$$= 20 + 28$$
$$= 48$$

d. $8 - |-3 \cdot 4 + 5| = 8 - |-12 + 5|$
$$= 8 - |-7|$$
$$= 8 - 7$$
$$= 1$$

33. a. $-89 + 57 + (-42) = -90 + 60 + (-40)$
$$= -30 + -40$$
$$= -70$$

b. $\dfrac{-507}{-24} = \dfrac{-500}{-25}$
$$= 20$$

c. $(-681)(9) = (-700)(10)$
$$= -7{,}000$$

d. $317 - (-775) = 300 + 800$
$$= 1{,}100$$

35. a. $t + (-8) = -18$
$$t + (-8) + 8 = -18 + 8$$
$$t = -10$$

b. $\dfrac{x}{-3} = -4$
$$-3 \cdot \dfrac{x}{-3} = -3 \cdot -4$$
$$x = 12$$

35. c. $y + 8 = -14$
$$y + 8 - 8 = -14 - 8$$
$$y = -22$$

d. $-7m = -28$
$$\dfrac{-7m}{-7} = \dfrac{-28}{-7}$$
$$m = 4$$

37. a.
$$-5t + 1 = -14$$
$$-5t + 1 - 1 = -14 - 1$$
$$-5t = -15$$
$$\frac{-5t}{-5} = \frac{-15}{-5}$$
$$t = 3$$

b.
$$6 = 2 - 2x$$
$$6 - 2 = 2 - 2x - 2$$
$$4 = -2x$$
$$\frac{4}{-2} = \frac{-2x}{-2}$$
$$-2 = x$$
$$x = -2$$

c.
$$\frac{x}{-4} - 5 = -1 - 1$$
$$\frac{x}{-4} - 5 = -2$$
$$\frac{x}{-4} - 5 + 5 = -2 + 5$$
$$\frac{x}{-4} = 3$$
$$-4\left(\frac{x}{-4}\right) = -4(3)$$
$$x = -12$$

d.
$$c - (-5) = 5$$
$$c + 5 = 5$$
$$c + 5 - 5 = 5 - 5$$
$$c = 0$$

39. Let x = the number of customers who applied for credit.
$$-8x = -968$$
$$\frac{-8x}{-8} = \frac{-968}{-8}$$
$$x = 121$$
121 customers applied for credit.

KEY CONCEPT: Signed Numbers

1. Stocks fell 5 points: -5

3. 30 seconds before going on the air: -30

5. 10 degrees above normal: $+10$

7. $205 overdrawn: -205

9.

11. $x < y$

13. **Addition**
Same sign: Add their absolute values and attach their common sign to the sum.
Different signs: Subtract their absolute values, the smaller from the larger, and attach the sign of the number with the larger absolute value to that result.

15. **Multiplication**
Same sign: The product is positive.
Different signs: The product is negative.

CHAPTER 2 TEST

1. **a.** $-8 \geq -9$

b. $-8 \leq |-8|$

c. The opposite of $5 \leq 0$

3. $a < 6$ yr

5. **a.** $-7 - 6 = -7 + (-6)$
$$= -13$$

b. $-7 - (-6) = -7 + 6$
$$= -1$$

7. $(-4)(5) = -20$

9. $\dfrac{-6 + (-10) + (-2)}{6} = \dfrac{-18}{6}$
$$= -3$$
Each will have to contribute $3 million.

11. **a.** $(-4)^2 = 16$
b. $-4^2 = -16$
c. $(-4 - 3)^2 = (-7)^2$
$$= 49$$

13. $4 - (-3)^2 + 6 = 4 - 9 + 6$
$$= 4 + (-9) + 6$$
$$= -5 + 6$$
$$= 1$$

15.

$$10 + 2[6 - (-2)(-5)] = 10 + 2(6 - 10)$$
$$= 10 + 2(-4)$$
$$= 10 + (-8)$$
$$= 2$$

17.

$$c - (-7) = -8$$
$$c + 7 = -8$$
$$c + 7 - 7 = -8 - 7$$
$$c = -8 + -7$$
$$c = -15$$

19.

$$\frac{x}{-4} = 10$$

$$-4\left(\frac{x}{-4}\right) = -4(10)$$

$$x = -40$$

21.

$$-5 = -6a + 7$$
$$-5 - 7 = -6a + 7 - 7$$
$$-5 + (-7) = -6a$$
$$-12 = -6a$$
$$\frac{-12}{-6} = \frac{-6a}{-6}$$
$$2 = a$$
$$a = 2$$

23. Let $x =$ the change in elevation

$$-650 + x = 400$$
$$-650 + x + 650 = 400 + 650$$
$$x = 1,050$$

They experienced a change of 1,050 ft.

CHAPTERS 1 & 2 CUMULATIVE REVIEW

1. Natural numbers: 1, 2, 5, 9

3. Negative numbers: –2, –1

5. 6 is in the thousands column.

7. 7,326,500

9. $237 + 549 = 786$

11.

$$\begin{array}{r} 5,369 \\ -\ \ 685 \\ \hline 4,684 \end{array}$$

13.

$$P = w + w + l + l$$
$$P = 17 + 17 + 35 + 35$$
$$P = 104$$

The perimeter is 104 ft.

15. $435 \cdot 27 = 11,745$

17.

$$\begin{array}{r} 4,587 \\ \times\ \ \ \ 67 \\ \hline 307,329 \end{array}$$

19.

$$A = l \cdot w$$
$$= 35 \cdot 17$$
$$= 595$$

The area is 595 ft^2.

21. 17 is prime and odd.

23. 0 is even.

25. $504 = 2^3 \cdot 3^2 \cdot 7$

27. $5^2 \cdot 7 = 25 \cdot 7$
$ = 175$

29. $25 + 5 \cdot 5 = 25 + 25$
$ = 50$

31. $\text{average} = \dfrac{38 + 42 + 36 + 38 + 48 + 44}{6}$

$\text{average} = \dfrac{246}{6}$

$\text{average} = 41$

On the average, they were not obeying the speed limit of 40 mph.

33. $\quad\quad 50 = x + 37$
$50 - 37 = x + 37 - 37$
$\quad\quad 13 = x$
$\quad\quad\ x = 13$
Check: $50 = x + 37$
$\quad\quad\quad\quad 50 \overset{?}{=} 13 + 37$
$\quad\quad\quad\quad 50 = 50$

35. $5p = 135$
$\dfrac{5p}{5} = \dfrac{135}{5}$
$\ p = 27$

37. $\quad 3n - 5 = 13$
$3n - 5 + 5 = 13 + 5$
$\quad\quad 3n = 18$
$\quad\quad \dfrac{3n}{3} = \dfrac{18}{3}$
$\quad\quad\ n = 6$

39.

41. $-2 + (-3) = -5$

43. $\left|-3\right| - 5 = 3 - 5$
$ = 3 + (-5)$
$ = -2$

45. $(-8)(-3) = 24$

47. $\dfrac{-14}{-7} = 2$

49. $5 + (-3)(-7) = 5 + 21$
$ = 26$

51. $\dfrac{10 - (-5)}{1 - 2 \cdot 3} = \dfrac{10 + 5}{1 - 6}$
$\phantom{\dfrac{10 - (-5)}{1 - 2 \cdot 3}} = \dfrac{15}{1 + (-6)}$
$\phantom{\dfrac{10 - (-5)}{1 - 2 \cdot 3}} = \dfrac{15}{-5}$
$\phantom{\dfrac{10 - (-5)}{1 - 2 \cdot 3}} = -3$

53. $\quad\quad -5t + 1 = -14$
$-5t + 1 - 1 = -14 - 1$
$\quad\quad\quad -5t = -15$
$\quad\quad\quad \dfrac{-5t}{-5} = \dfrac{-15}{-5}$
$\quad\quad\quad\ t = 3$

55. Let $x =$ the amount of each person's share
$$12x = -1,512,444$$
$$\frac{12x}{12} = \frac{-1,512,444}{12}$$
$$x = -126,037$$
Each person's share was $126,037.

STUDY SET Section 3.1

VOCABULARY

1. An <u>algebraic expression</u> is a combination of variables, numbers, and the operation symbols for addition, subtraction, multiplication and division.

3. A <u>variable</u> is a letter that is used to stand for a number. A <u>constant</u> is a number that is fixed and does not change its value.

CONCEPTS

5. $10 + 3x$; $\dfrac{10 - x}{3}$

   (answers may vary)

7. **a.** Lopez: d mi ;
   Lamb: $d + 15$ mi

   b. Lamb lives 15 miles farther.

9.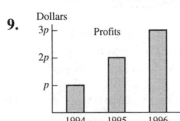

11.

Wind conditions	Speed of jet
In still air	500
With the wind	$500 + x$
Against the wind	$500 - x$

13. p = the cost of parts;
    $400 - p$ = labor cost

15. She should study $\dfrac{h}{4}$ hours each day.

NOTATION

17. $8x$

19. $\dfrac{10}{c}$

PRACTICE

21. $x - 9$

23. $\dfrac{2}{3}$

25. $6 + r$

27. $d - 15$

29. $1 - s$

31. $2p$

33. $s + 14$

35. $\dfrac{35}{b}$

37. $x - 2$

39. c increased by 7

41. 7 less than c

43. **a.** $60m$ seconds in m minutes

    b. $3,600h$ seconds in h hours

45. **a.** Her salary per month is $\dfrac{s}{12}$.

    b. Her salary per week is $\dfrac{s}{52}$.

47. **a.** Its length in inches is $12f$.

47. b. Its length in yards is $\dfrac{f}{3}$.

49. $j - 5$

51. $6s$

53. $\dfrac{p}{15}$

55. $t + 2$

57. w = width; $w + 6$ = length

59. g = number of gallons drained out; $6 - g$ = number of gallons remaining

61. $3x + 5$

63. $10a + 12$

65. $70 - 5d$

67. $4x + 3$

69. $3,000 - 50x$

APPLICATIONS

71. x = number of votes received by Nixon; $x + 118,550$ = number of votes received by Kennedy

73. x = length of the shortest strip; $x + 6$ = length of middle strip; $x + 12$ = length of longest strip

REVIEW

79. $-5 + (-6) + 1 = -11 + 1$
$$= -10$$

81.
$$-x = 4$$
$$-1x = 4$$
$$-1(-1x) = -1(4)$$
$$x = -4$$

STUDY SET Section 3.2

VOCABULARY

1. A <u>formula</u> is a mathematical expression that states a known relationship between two or more variables.

3. To evaluate an algebraic expression, we <u>substitute</u> specific numbers for the variables in the expression and apply the rules for order of operations.

CONCEPTS

5. $2 - 8 + 10$ makes it look like a subtraction.

7. x = length of part 1; $x - 40$ = length of part 2; $x + 16$ = length of part 3

9.

Ticket price	Service charge	Total cost
$20	$2	**$22**
$25	$2	**$27**
$p	$2	**$(p + 2)**
$10p	$2	**$(10p + 2)**

11.

	Rate (mph)	Time (hr)	Distance traveled
Bus	25	2	50 mi
Bike	12	4	48 mi
Walk	3	t	$3t$ mi
Run	5	1	5 mi
Car	x	3	$3x$ mi

13. speedometer: rate; odometer: distance; clock: time; $d = rt$

15. The rate is expressed in miles per hour and the time is in minutes.

NOTATION

17. a. $d = rt$

b. $C = \dfrac{5(F - 32)}{9}$

c. $d = 16t^2$

PRACTICE

19. $3x + 5 = 3(4) + 5$
$= 12 + 5$
$= 17$

21. $-p = -(-4)$
$= 4$

23. $-4t = -4(-10)$
$= 40$

25. $\dfrac{x - 8}{2} = \dfrac{-4 - 8}{2}$
$= \dfrac{-12}{2}$
$= -6$

27. $2(p + 9) = 2(-12 + 9)$
$= 2(-3)$
$= -6$

29. $x^2 + 14 = 3^2 + 14$
$= 9 + 14$
$= 23$

31. $8s - s^3 = 8(-2) - (-2)^3$
$= 8(-2) - (-8)$
$= -16 - (-8)$
$= -8$

33. $4x^2 = 4(5)^2$
$= 4(25)$
$= 100$

35. $3b - b^2 = 3(-4) - (-4)^2$
$= 3(-4) - 16$
$= -12 - 16$
$= -28$

37. $\dfrac{24 + k}{3k} = \dfrac{24 + 3}{3(3)}$
$= \dfrac{27}{9}$
$= 3$

39. $\dfrac{x}{y} = \dfrac{30}{-10}$
$= -3$

41. $-x - y = -(-1) - 8$
$= 1 - 8$
$= -7$

43. $x(5h - 1) = -2[5(2) - 1]$
$= -2(10 - 1)$
$= -2(9)$
$= -18$

45. $b^2 - 4ac = (-3)^2 - 4(4)(-1)$
$= 9 - 4(4)(-1)$
$= 9 - (-16)$
$= 25$

47. $x^2 - y^2 = 5^2 - (-2)^2$
$= 25 - 4$
$= 21$

49. $\dfrac{50 - 6s}{-t} = \dfrac{50 - 6(5)}{-4}$
$= \dfrac{50 - 30}{-4}$
$= \dfrac{20}{-4}$
$= -5$

51. $-5abc + 1 = -5(-2)(-1)(3) + 1$
$= -30 + 1$
$= -29$

53. $5s^2 t = 5(-3)^2 (-1)$
$= 5(9)(-1)$
$= -45$

55. $r = c + m$
$r = 20 + 50$
$r = 70$
The price is 70 cents.

57. $p = r - c$
$p = 13,500 - 5,300$
$p = 8,200$
The profit was \$8,200.

59. $r = c + m$
$r = 18 + 5$
$r = 23$
The price is \$23.

61. $d = rt$
$d = 60(5)$
$d = 300$
The distance is 300 miles.

63. $C = \dfrac{5(F - 32)}{9}$
$C = \dfrac{5(14 - 32)}{9}$
$= \dfrac{5(-18)}{9}$
$= \dfrac{-90}{9}$
$C = -10$
The temperature is $-10°C$.

65. $A = \dfrac{S}{n}$
$A = \dfrac{254 + 225 + 238}{3}$
$= \dfrac{717}{3}$
$A = 239$
The average is 239.

67. $d = 16t^2$
$d = 16(2)^2$
$= 16(4)$
$d = 64$
The ball has fallen 64 ft.

APPLICATIONS

69. a. 1993: $p = 10 - 10$ so they broke even
1994: $p = 20 - 25$ so they lost \$5 million
1995: $p = 20 - 15$ so they had a \$5 million profit

69. b.

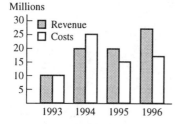

Millions

1996: $p = 25 - 15$ so they
had a $10 million profit.

71. $D = B - C$

	A	B	C	D
1	Bath towel set	$25	$5	**$20**
2	Pillows	$15	$3	**$12**
3	Comforter	$53	$11	**$42**

73. $A = \dfrac{S}{n}$

$$A = \frac{16 + 10 + 4 + (-2) + 0 + (-4)}{6}$$

$$= \frac{24}{6}$$

$$A = 4$$

His average was 4 yards per carry.

CALCULATOR

75. $A = \dfrac{S}{n}$

$$A = \frac{35 + 50 + 60 + 55 + 40 + 25 + 15}{7}$$

$$= \frac{280}{7}$$

$$A = 40$$

The average for that week was 40 ships per day.

77. $A = \dfrac{S}{n}$

$$A = \frac{53(5) + 26(3) + 9(1)}{53 + 26 + 9}$$

$$A = 4$$

The average score was 4.

REVIEW

83. 17, 37, and 41 are prime.

85. $\left| -2 + (-5) \right| = \left| -7 \right|$

$$= 7$$

87. Division by 3 is performed on the variable.

89. $-3 - (-6) = -3 + 6$

$$= 3$$

STUDY SET Section 3.3

VOCABULARY

1. The <u>distributive property</u> tells us how to multiply $5(7 + x)$. After doing the multiplication to obtain $35 + 5x$, we say that the parentheses were <u>removed</u>.

3. When an algebraic expression is simplified, the result is an <u>equivalent</u> expression.

CONCEPTS

5. $x(y + z) = xy + xz$

7. $(w + 7)5$ is the right distributive property.

9. Two groups of 6 plus three groups of 6 is <u>5</u> groups of 6. Therefore,
$6 \cdot 2 + \underline{6} \cdot 3 = 6(\underline{2} + \underline{3})$.

11. $-(y + 9) = -y - 9$

13. The arrows show distributing the -9.

NOTATION

15. $-5(7n) = (\underline{-5} \cdot 7)n$
$$= -35n$$

17. $-9(-4 - 5y) = (\underline{-9})(-4) - \left(\underline{-9}\right)(5y)$
$$= 36 - (\underline{-45y})$$
$$= 36 + 45y$$

PRACTICE

19. $2(6x) = (2 \cdot 6)x$
$$= 12x$$

21. $-5(6y) = (-5 \cdot 6)y$
$$= -30y$$

23. $-10(-10t) = (-10 \cdot -10)t$
$$= 100t$$

25. $(4s)3 = (4 \cdot 3)s$
$$= 12s$$

27. $2c \cdot 7 = (2 \cdot 7)c$
$$= 14c$$

29. $-5 \cdot 8h = (-5 \cdot 8)h$
$$= -40h$$

31. $-7x(6y) = [-7(6)](x \cdot y)$
$$= -42xy$$

33. $4r \cdot 4s = [4(4)](r \cdot s)$
$$= 16rs$$

35. $2x(5y)(3) = [2(5)(3)](x \cdot y)$
$$= 30xy$$

37. $5r(2)(-3b) = [5(2)(-3)](r \cdot b)$
$$= -30br$$

39. $5 \cdot 8c \cdot 2 = [5(8)(2)]c$
$$= 80c$$

41. $(-1)(-2e)(-4) = [(-1)(-2)(-4)]e$
$$= -8e$$

43. $4(x + 1) = 4x + 4(1)$
$$= 4x + 4$$

45. $4(4 - x) = 4(4) - 4x$
$$= 16 - 4x$$

47. $-2(3e + 3) = -2(3e) + (-2)(3)$
$$= -6e + (-6)$$
$$= -6e - 6$$

49. $-8(2q - 6) = -8(2q) - (-8)(6)$
$$= -16q - (-48)$$
$$= -16q + [-(-48)]$$
$$= -16q + 48$$

51.
$$-4(-3-5s)=-4(-3)-(-4)(5s)$$
$$=12-(-20s)$$
$$=12+[-(-20s)]$$
$$=12+20s$$

53.
$$(7+4d)6=(7)6+(4d)6$$
$$=42+24d$$

55.
$$(5r-6)(-5)=(5r)(-5)-(6)(-5)$$
$$=-25r-(-30)$$
$$=-25r+[-(-30)]$$
$$=-25r+30$$

57.
$$(-4-3d)6=(-4)6-(3d)6$$
$$=-24-18d$$

59.
$$3(3x-7y+2)=3(3x)-3(7y)+3(2)$$
$$=9x-21y+6$$

61.
$$-3(-3z-3x-5y)=-3(-3z)-(-3)(3x)-(-3)(5y)$$
$$=9z-(-9x)-(-15y)$$
$$=9z+[-(-9x)]+[-(-15y)]$$
$$=9z+9x+15y$$

63.
$$-(x+3)=-1(x+3)$$
$$=-1(x)+(-1)(3)$$
$$=-x+-3$$
$$=-x-3$$

65.
$$-(4t+5)=-1(4t+5)$$
$$=-1(4t)+(-1)(5)$$
$$=-4t+(-5)$$
$$=-4t-5$$

67.
$$-(-3w-4)=-1(-3w-4)$$
$$=-1(-3w)-(-1)(4)$$
$$=3w-(-4)$$
$$=3w+[-(-4)]$$
$$=3w+4$$

69. $-(5x - 4y + 1) = -1(5x - 4y + 1)$
$$= -1(5x) - (-1)(4y) + (-1)(1)$$
$$= -5x - (-4y) + (-1)$$
$$= -5x + [-(-4y)] + (-1)$$
$$= -5x + 4y - 1$$

70. $-(6r - 5f + 1) = -1(6r - 5f + 1)$
$$= -1(6r) - (-1)(5f) + (-1)(1)$$
$$= -6r - (-5f) + (-1)$$
$$= -6r + [-(-5f)] + (-1)$$
$$= -6r + 5f - 1$$

71. $2(5) + 2(4x) = 2(5 + 4x)$

73. $-4(5) - 3x(5) = (-4 - 3x)5$

75. $-3(4y) - (-3)(2) = -3(4y - 2)$

77. $3(4) - 3(7t) - 3(5s) = 3(4 - 7t - 5s)$

REVIEW

83. analyze, form, solve, state, check

85. product : <u>multiplication</u>; quotient: <u>division</u>; difference: <u>subtraction</u>; sum: <u>addition</u>

87. $-6 \geq -7$

89. Carpeting a room and painting a wall involve area.

STUDY SET Section 3.4

VOCABULARY

1. A <u>term</u> is a number or a product of a number and one or more variables.

3. The <u>perimeter</u> of a geometric figure is the distance around it.

5. $2(x + 3) = 2x + 2 \cdot 3$ is an example of the <u>distributive</u> property.

7. When an algebraic expression contains like terms, an <u>equivalent</u> expression can be obtained by <u>combining</u> like terms.

CONCEPTS

9. a. In $12 + x$, x is used as a term.

   b. In $7x$, x is used as a factor.

   c. In $12y + 12x - 6$, x is used as a factor.

   d. In $-36xy$, x is used as a factor.

11. $a(b + c) = ab + ac$

13. a. 11

13. b. 8

c. -4

d. 1

e. -1

f. 102

15. It helps identify the like terms.

17. $(d + 15) + d = (2d + 15)$ miles

19. To combine like terms, add the coefficients of the like terms.

NOTATION

21. $5x + 7x = (5 + \underline{7})x$
$= 12x$

23. $2a - 3b + 5a = 2a + 5a - \underline{3b}$
$= (2 + 5)a - \underline{3b}$
$= 7a - 3b$

25. $2(x - y) + 3x = 2x - \underline{2y} + 3x$
$= 5x - 2y$

PRACTICE

27. $6t + 9t = 15t$

29. $5s - s = 4s$

31. $-5x + 6x = x$

33. $-5d + 9d = 4d$

35. $3e - 7e = -4e$

37. $-3x - 4x = -7x$

39. $4z - 10z = -6z$

41. $h - 7b$ cannot be simplified.

43. In $3x^2 - 5x + 4$ the terms are $3x^2$, $-5x$, and 4.

45. In $5 + 5t - 8t + 4$, the terms are 5, $5t$, $-8t$, and 4.

47. $3x^2y, -6x^2y$

49. $-8h^5c^2, -5h^5c^2$

51. $6t + 9 + 5t + 3 = 6t + 5t + 9 + 3$
$= 11t + 12$

53. $3w - 4 - w - 1 = 3w - w - 4 - 1$
$= 2w - 5$

55. $-4r + 8R + 2R - 3r + R = -7r + 11R$

57. $-45d - 12a - 5d + 12a = -50d$

59. $4x - 3y - 7 + 4x - 2 - y = 8x - 4y - 9$

61. $4(x + 1) + 5(6 + x) = 4x + 4 + 30 + 5x$
$= 9x + 34$

63. $5(3-2s)+4(2-3s)=15-10s+8-12s$
$$=-22s+23$$

65. $-4(6-4e)+3(e+1)=-24+16e+3e+3$
$$=19e-21$$

67. $3t-(t-8)=3t-t+8$
$$=2t+8$$

69. $-2(2-3x)-3(x-4)=-4+6x-3x+12$
$$=3x+8$$

71. $-4(-4y+5)-6(y+2)=16y-20-6y-12$
$$=10y-32$$

APPLICATIONS

73. perimeter = 2(perimeter of front) + 2(perimeter of side)
$$P=2(2l+2w)+2(2l+2w)$$
$$=2(2\cdot 60+2\cdot 10)+2(2\cdot 10+2\cdot 10)$$
$$=2(120+20)+2(20+20)$$
$$=2(140)+2(40)$$
$$=280+80$$
$$=360$$
360 ft of pine is needed.

REVIEW

79.
$$-4t-3=-11$$
$$-4t-3+3=-11+3$$
$$-4t=-8$$
$$\frac{-4t}{-4}=\frac{-8}{-4}$$
$$t=2$$

81. $100=2^2\cdot 5^2$

83. The <u>absolute value</u> of a number is the distance between it and 0 on the number line.

STUDY SET Section 3.5

VOCABULARY

1. To <u>solve</u> an equation means to find all values of the variable that make the equation a true statement.

3. In $2(x+4)$, to remove parentheses means to apply the <u>distributive</u> property.

5. The phrase *combine like terms* refers to the operations of <u>addition</u> and <u>subtraction</u>.

CONCEPTS

7. The addition property of equality

9. The subtraction property of equality

11. a.
$$2(x+1) = 2(-4+1)$$
$$= 2(-3)$$
$$= -6$$

b.
$$2(x+1) = 2x + 2(1)$$
$$= 2x + 2$$

c.
$$2(x+1) = -4$$
$$2x + 2 = -4$$
$$2x + 2 - 2 = -4 - 2$$
$$2x = -6$$
$$\frac{2x}{2} = \frac{-6}{2}$$
$$x = -3$$

13. a.
$$3x + 4 - x = 2x + 4$$

b.
$$3x + 4 - x = 8$$
$$2x + 4 = 8$$
$$2x + 4 - 4 = 8 - 4$$
$$2x = 4$$
$$\frac{2x}{2} = \frac{4}{2}$$
$$x = 2$$

NOTATION

15.
$$4x + 5 - 2x = -15$$
$$2x + 5 = -15$$
$$2x + 5 - 5 = -15 - 5$$
$$2x = -20$$
$$\frac{2x}{2} = \frac{-20}{2}$$
$$x = -10$$

17.
$$5(x - 9) = 5$$
$$5x - 5(9) = 5$$
$$5x - 45 = 5$$
$$5x - 45 + 45 = 5 + 45$$
$$5x = 50$$
$$\frac{5x}{5} = \frac{50}{5}$$
$$x = 10$$

PRACTICE

19.
$$5y - 7 + 1 = 10 - 1$$
$$5y - 6 = 9$$
$$5y - 6 + 6 = 9 + 6$$
$$5y = 15$$
$$\frac{5y}{5} = \frac{15}{5}$$
$$y = 3$$

21.
$$5a - 4 + a = -28$$
$$6a - 4 = -28$$
$$6a - 4 + 4 = -28 + 4$$
$$6a = -24$$
$$\frac{6a}{6} = \frac{-24}{6}$$
$$a = -4$$

23.
$$-3x - 5 = -7 - 16$$
$$-3x - 5 = -23$$
$$-3x - 5 + 5 = -23 + 5$$
$$-3x = -18$$
$$\frac{-3x}{-3} = \frac{-18}{-3}$$
$$x = 6$$

25.
$$4(x-2)=0$$
$$4x-8=0$$
$$4x-8+8=0+8$$
$$4x=8$$
$$\frac{4x}{4}=\frac{8}{4}$$
$$x=2$$

27.
$$3s+1=4s-7$$
$$3s+1-3s=4s-7-3s$$
$$1=s-7$$
$$1+7=s-7+7$$
$$8=s$$

29.
$$38-5w=-10+7w$$
$$38-5w+5w=-10+7w+5w$$
$$38=-10+12w$$
$$38+10=-10+12w+10$$
$$48=12w$$
$$\frac{48}{12}=\frac{12w}{12}$$
$$4=w$$

31.
$$2(x+6)=4$$
$$2x+12=4$$
$$2x+12-12=4-12$$
$$2x=-8$$
$$\frac{2x}{2}=\frac{-8}{2}$$
$$x=-4$$

33.
$$-(c-4)=3$$
$$-c+4=3$$
$$-c+4-4=3-4$$
$$-c=-1$$
$$-1(-c)=-1(-1)$$
$$c=1$$

35.
$$-3(2w-3)=9$$
$$-6w+9=9$$
$$-6w+9-9=9-9$$
$$-6w=0$$
$$\frac{-6w}{-6}=\frac{0}{-6}$$
$$w=0$$

37.
$$-16=2(t+4)$$
$$-16=2t+8$$
$$-16-8=2t+8-8$$
$$-24=2t$$
$$\frac{-24}{2}=\frac{2t}{2}$$
$$-12=t$$

39.
$$\frac{x}{2}-2-1=4$$
$$\frac{x}{2}-3=4$$
$$\frac{x}{2}-3+3=4+3$$
$$\frac{x}{2}=7$$
$$2\left(\frac{x}{2}\right)=2(7)$$
$$x=14$$

41.
$$1-3-5+\frac{c}{-4}=0$$
$$-7+\frac{c}{-4}=0$$
$$-7+\frac{c}{-4}+7=0+7$$
$$\frac{c}{-4}=7$$
$$-4\left(\frac{c}{-4}\right)=-4(7)$$
$$c=-28$$

43.
$$2(x+6)-4=2$$
$$2x+12-4=2$$
$$2x+8=2$$
$$2x+8-8=2-8$$
$$2x=-6$$
$$\frac{2x}{2}=\frac{-6}{2}$$
$$x=-3$$

45.
$$2(4y+8)=3(2y-2)$$
$$8y+16=6y-6$$
$$8y+16-6y=6y-6-6y$$
$$2y+16=-6$$
$$2y+16-16=-6-16$$
$$2y=-22$$
$$\frac{2y}{2}=\frac{-22}{2}$$
$$y=-11$$

47.
$$5-(7-y)=-5$$
$$5-7+y=-5$$
$$-2+y=-5$$
$$-2+y+2=-5+2$$
$$y=-3$$

49.
$$4(r-1)-5(2r+6)=-4$$
$$4r-4-10r-30=-4$$
$$-6r-34=-4$$
$$-6r-34+34=-4+34$$
$$-6r=30$$
$$\frac{-6r}{-6}=\frac{30}{-6}$$
$$r=-5$$

51.
$$-(6x+3)+9(2x+2)=3$$
$$-6x-3+18x+18=3$$
$$12x+15=3$$
$$12x+15-15=3-15$$
$$12x=-12$$
$$\frac{12x}{12}=\frac{-12}{12}$$
$$x=-1$$

53.
$$-3(4-2x)-5(2x+1)=-1$$
$$-12+6x-10x-5=-1$$
$$-4x-17=-1$$
$$-4x-17+17=-1+17$$
$$-4x=16$$
$$\frac{-4x}{-4}=\frac{16}{-4}$$
$$x=-4$$

55.
$$35p+2-(15p+3)=58+1$$
$$35p+2-15p-3=59$$
$$20p-1=59$$
$$20p-1+1=59+1$$
$$20p=60$$
$$\frac{20p}{20}=\frac{60}{20}$$
$$p=3$$

57.
$$4x+2(x-1)=-33$$
$$4(-5)+2(-5-1)\overset{?}{=}-33$$
$$4(-5)+2(-6)\overset{?}{=}-33$$
$$-20+-12\overset{?}{=}-33$$
$$-32\neq-33$$

No, –5 is not a solution.

59. $6f + 8 - f = 11 + 4f$

$6(3) + 8 - 3 \overset{?}{=} 11 + 4(3)$

$18 + 8 - 3 \overset{?}{=} 11 + 12$

$23 = 23$

Yes, 3 is a solution.

REVIEW

65. $-7 - 9 = -16$

67. $\dfrac{-8 + 2}{-2 + 4} = \dfrac{-6}{2}$

$= -3$

69. $-(-5) = 5$

71. The product of two negative integers will be positive.

STUDY SET Section 3.6

VOCABULARY

1. The words *increased by*, *longer*, *taller*, *higher*, and *more than* indicate that the operation of <u>addition</u> should be used.

CONCEPTS

3. Step 1: Analyze the problem
Step 2: Form an equation
Step 3: Solve the equation
Step 4: State the conclusion
Step 5: Check the result

5. He owes $35p$ in parking fines.

7. Let x = one number.
$x - 5$ = the other number

9. a. $p - 50$ people preferred the fat-free margarine.

b. $2p - 50$ people participated in the taste test.

11.

Denomination of savings bond	Number	Total value
$25	10	**$250**
$50	f	**$50f**
$100	d	**$100d**
$500	x	**$500x**

13. a. In solving the equation $2x = 10$, we undo the <u>multiplication</u> of the variable by 2 by <u>dividing</u> both sides of the equation by 2.

b. In solving the equation $x + 2 = 10$, we undo the <u>addition</u> of 2 to the variable by <u>subtracting</u> 2 from both sides of the equation.

NOTATION

15. A: You are asked to find <u>the number of economy and first-class seats</u>.
There are <u>2</u> unknowns.

   F: Since the number of economy seats is related to the number of <u>first-class seats</u>, we will let
$x =$ <u>the number of first-class seats</u>
Key phrase: <u>ten times as many</u>
Translation: <u>multiply by 10</u>
So $10x =$ the number of economy seats

The number of first-class seats	plus	the number of economy seats	is	88.
x	$+$	$10x$	$=$	88

   S: $x + 10x = 88$
$$11x = 88$$
$$x = 8$$

   S: There are <u>8</u> first-class seats and <u>80</u> economy seats.

   C: If we add <u>8</u> and <u>80</u>, we get 88. The answers check.

APPLICATIONS

17. Let $x =$ the number of months it would take to reach his goal.
$$15 + 5x = 100$$
$$15 + 5x - 15 = 100 - 15$$
$$5x = 85$$
$$x = 17$$
It would take him 17 months.

19. Let $x =$ the amount the premium tank holds.
$x - 100 =$ the amount the regular tank holds.
$$x + x - 100 = 700$$
$$2x - 100 = 700$$
$$2x = 800$$
$$x = 400$$
The premium tank holds 400 gallons.

21. Let $x =$ the distance the freighter was from port.
$3x =$ the distance the passenger ship was from port.
$$x + 3x = 84$$
$$4x = 84$$
$$x = 21$$
The freighter was 21 miles from port.

23. Let $w =$ the width of the room.
$2w =$ the length of the room.
$$2 \cdot 2w + 2w = 60$$
$$4w + 2w = 60$$
$$6w = 60$$
$$w = 10$$
The width of the room is 10 ft.

25. Let $x =$ the number of minutes of commercials.
$x + 18 =$ the number of minutes for the program.
$$x + x + 18 = 30$$
$$2x + 18 = 30$$
$$2x = 12$$
$$x = 6$$
There were 6 minutes of commercials.

27. Let x = the amount of the monthly rent.

$$100 + \frac{x}{3} = 225$$

$$\frac{x}{3} = 125$$

$$x = 375$$

The monthly rent was $375.

29. Let x = the number of offices to which he delivered.

$$300 - 3x = 117$$

$$-3x = -183$$

$$x = 61$$

He delivered to 61 offices.

31. Let x = the number of dress shoes sold.
$9 - x$ = the number of athletic shoes sold.

$$3x + 2(9 - x) = 24$$

$$3x + 18 - 2x = 24$$

$$x + 18 = 24$$

$$x = 6$$

The student had 8 correct answer and 2 incorrect answers.

33. Let x = the number of correct answers
$10 - x$ = the number of incorrect answers

$$3x - 4(10 - x) = 16$$

$$3x - 40 + 4x = 16$$

$$7x - 40 = 16$$

$$7x = 56$$

$$x = 8$$

The student had 8 correct answers and 2 incorrect answers.

REVIEW

39. The associative property of addition

41. $-10^2 = -(10 \cdot 10)$
$= -100$

43. Subtraction of a number is the same as <u>addition</u> of the opposite of that number.

45. $2 \cdot 2 \cdot 2 \cdot 5 \cdot 5 = 2^3 \cdot 5^2$

STUDY SET Section 3.7

VOCABULARY

1. In x^n, x is called the <u>base</u> and n is called the <u>exponent</u>.

3. $x^m \cdot x^n$ is the <u>product</u> of two exponential expressions with <u>like</u> bases.

5. $(2x)^n$ is a <u>product</u> raised to a power.

CONCEPTS

7. a. $x \cdot x \cdot x \cdot x \cdot x \cdot x \cdot x = x^7$

b. $x \cdot x \cdot y \cdot y \cdot y = x^2 y^3$

c. $3 \cdot 3 \cdot 3 \cdot 3 \cdot a \cdot a \cdot b \cdot b \cdot b = 3^4 a^2 b^3$

9. $x^2 \cdot x^6 = x^8$ (answers may vary)

11. $(c^5)^2 = c^{10}$ (answers may vary)

13. a. $x^m x^n = x^{m+n}$

b. $(x^m)^n = x^{mn}$

c. $x^2 \cdot x^2 = x^4$; $x^2 + x^2 = 2x^2$

15. a. $2^1 = 2$

b. $(-10)^1 = -10$

c. $x^1 = x$

17. a. $x \cdot x = x^2$; $x + x = 2x$

b. $x \cdot x^2 = x^3$; $x + x^2 = x + x^2$

c. $x^2 \cdot x^2 = x^4$; $x^2 + x^2 = 2x^2$

19. a. $4x \cdot x = 4x^2$; $4x + x = 5x$

b. $4x \cdot 3x = 12x^2$; $4x + 3x = 7x$

c. $4x^2 \cdot 3x = 12x^3$;
$4x^2 + 3x = 4x^2 + 3x$

21. $x^{m+n} = 3^{2+1}$
$= 3^3$
$= 27$

NOTATION

23. $x^5 \cdot x^7 = x^{5+7}$
$= x^{12}$

25. $(2x^4)(8x^3) = (2 \cdot 8)(x^4 \cdot x^3)$
$= 16x^{4+3}$
$= 16x^7$

PRACTICE

27. $x^2 \cdot x^3 = x^{2+3}$
$= x^5$

29. $x^3 x^7 = x^{3+7}$
$= x^{10}$

31. $f^5(f^8) = f^{5+8}$
$= f^{13}$

33. $n^{24} \cdot n^8 = n^{24+8}$
$= n^{32}$

35. $l^4 \cdot l^5 \cdot l = l^{4+5+1}$
$= l^{10}$

37. $x^6(x^3)x^2 = x^{6+3+2}$
$= x^{11}$

39. $2^4 \cdot 2^8 = 2^{4+8}$
$= 2^{12}$

41. $5^6(5^2) = 5^{6+2}$
$= 5^8$

43. $2x^2 \cdot 4x = (2 \cdot 4)(x^2 \cdot x)$
$= 8x^{2+1}$
$= 8x^3$

45. $5t \cdot t^9 = 5(t \cdot t^9)$
$= 5t^{1+9}$
$= 5t^{10}$

47. $-6x^3(4x^2) = (-6 \cdot 4)(x^3 \cdot x^2)$
$= -24x^{3+2}$
$= -24x^5$

49. $-x \cdot x^3 = -1(x \cdot x^3)$
$$= -x^{1+3}$$
$$= -x^4$$

51. $6y(2y^3)3y^4 = (6 \cdot 2 \cdot 3)(y \cdot y^3 \cdot y^4)$
$$= 36y^{1+3+4}$$
$$= 36y^8$$

53. $-2t^3(-4t^2)(-5t^5) = (-2 \cdot -4 \cdot -5)(t^3 \cdot t^2 \cdot t^5)$
$$= -40t^{3+2+5}$$
$$= -40t^{10}$$

55. $xy^2 \cdot x^2y = (x \cdot x^2)(y^2 \cdot y)$
$$= x^{1+2}y^{2+1}$$
$$= x^3y^3$$

57. $b^3 \cdot c^2 \cdot b^5 \cdot c^6 = (b^3 \cdot b^5)(c^2 \cdot c^6)$
$$= b^{3+5}c^{2+6}$$
$$= b^8c^8$$

59. $x^4y(xy) = (x^4 \cdot x)(y \cdot y)$
$$= x^{4+1}y^{1+1}$$
$$= x^5y^2$$

61. $a^2b \cdot b^3a^2 = (a^2 \cdot a^2)(b \cdot b^3)$
$$= a^{2+2}b^{1+3}$$
$$= a^4b^4$$

63. $x^5y \cdot y^6 = x^5(y \cdot y^6)$
$$= x^5y^{1+6}$$
$$= x^5y^7$$

65. $3x^2y^3 \cdot 6xy = (3 \cdot 6)(x^2 \cdot x)(y^3 \cdot y)$
$$= 18x^{2+1}y^{3+1}$$
$$= 18x^3y^4$$

67. $xy^2 \cdot 16x^3 = 16(x \cdot x^3)y^2$

$\qquad = 16x^{1+3}y^2$

$\qquad = 16x^4y^2$

69. $-6f^2t(4f^4t^3) = (-6 \cdot 4)(f^2 \cdot f^4)(t \cdot t^3)$

$\qquad = -24f^{2+4}t^{1+3}$

$\qquad = -24f^6t^4$

71. $ab \cdot ba \cdot a^2b = (a \cdot a \cdot a^2)(b \cdot b \cdot b)$

$\qquad = a^{1+1+2}b^{1+1+1}$

$\qquad = a^4b^3$

73. $-4x^2y(-3x^2y^2) = (-4 \cdot -3)(x^2 \cdot x^2)(y \cdot y^2)$

$\qquad = 12x^{2+2}y^{1+2}$

$\qquad = 12x^4y^3$

75. $(x^2)^4 = x^{2 \cdot 4}$

$\qquad = x^8$

77. $(m^{50})^{10} = m^{50 \cdot 10}$

$\qquad = m^{500}$

79. $(2a)^3 = 2^3a^3$

$\qquad = 8a^3$

81. $(xy)^4 = x^4y^4$

83. $(3s^2)^3 = 3^3 \cdot s^{2 \cdot 3}$

$\qquad = 27s^6$

85. $(2s^2t^3)^2 = 2^2s^{2 \cdot 2}t^{3 \cdot 2}$

$\qquad = 4s^4t^6$

87. $(x^2)^3(x^4)^2 = x^{2 \cdot 3}x^{4 \cdot 2}$

$\qquad = x^6x^8$

$\qquad = x^{6+8}$

$\qquad = x^{14}$

89. $(c^5)^3(c^3)^5 = c^{5 \cdot 3} \cdot c^{3 \cdot 5}$

$\qquad = c^{15} \cdot c^{15}$

$\qquad = c^{15+15}$

$\qquad = c^{30}$

91. $(2a^4)^2(3a^3)^2 = 2^2(a^4)^2 \cdot 3^2(a^3)^2$

$\qquad = 2^2a^{4 \cdot 2} \cdot 3^2a^{3 \cdot 2}$

$\qquad = 2^2a^8 \cdot 3^2a^6$

$\qquad = (2^2 \cdot 3^2)(a^8 \cdot a^6)$

$\qquad = (4 \cdot 9)a^{8+6}$

$\qquad = 36a^{14}$

93. $(3a^3)^3(2a^2)^3 = 3^3(a^3)^3 \cdot 2^3(a^2)^3$
$$= 3^3 a^{3\cdot3} \cdot 2^3 a^{2\cdot3}$$
$$= 3^3 a^9 \cdot 2^3 a^6$$
$$= (3^3 \cdot 2^3)(a^9 \cdot a^6)$$
$$= (27 \cdot 8)a^{9+6}$$
$$= 216a^{15}$$

95. $\left(x^2 x^3\right)^{12} = \left(x^{2+3}\right)^{12}$
$$= \left(x^5\right)^{12}$$
$$= x^{5\cdot12}$$
$$= x^{60}$$

97. $\left(2b^4 b\right)^5 = \left(2b^{4+1}\right)^5$
$$= \left(2b^5\right)^5$$
$$= 2^5\left(b^5\right)^5$$
$$= 2^5 b^{5\cdot5}$$
$$= 32b^{25}$$

REVIEW

103. A variable is a letter used to represent a number.

105. $\dfrac{-25}{-5} = 5$

107. $2\left(\dfrac{12}{-3}\right) + 3(5) = 2(-4) + 3(5)$
$$= -8 + 15$$
$$= 7$$

109. $-x = -12$
$$-1(-x) = -1(-12)$$
$$x = 12$$

CHAPTER 3 REVIEW

1. a. Brandon is closer by 250 mi.

   b. The height of the ceiling is $(h + 7)$ ft.

3. a. $\dfrac{c}{6}$

   b. $1,000 - 50w$

5. a. x dozen has $12x$ eggs.

   b. d days is $\dfrac{d}{7}$ weeks.

7. a. $-2x + 6 = -2(-3) + 6$
$$= 6 + 6$$
$$= 12$$

   b. $\dfrac{6-a}{1+a} = \dfrac{6-(-2)}{1+(-2)}$
$$= \dfrac{8}{-1}$$
$$= -8$$

   c. $b^2 - 4ac = 6^2 - 4(4)(-4)$
$$= 36 - (-64)$$
$$= 36 + 64$$
$$= 100$$

   d. $\dfrac{-2k^3}{1-2-3} = \dfrac{-2(-2)^3}{1-2-3}$
$$= \dfrac{-2(-8)}{1+(-2)+(-3)}$$
$$= \dfrac{16}{-1+(-3)}$$
$$= \dfrac{16}{-4}$$
$$= -4$$

9. $s = p - d$
$s = 315 - 37$
$= 278$
The sale price is \$278.

11. a. 1994 had the most revenue.

    b. $p = r - c$; 1996 had the most profit.

    c. The costs decreased over the 3-year span.

13. $d = 16t^2$
$= 16(3)^2$
$= 16(9)$
$= 144$
In 3 seconds the hammer will fall 144 ft.

15. a. $-2(5x) = (-2 \cdot 5)x$
$= -10x$

15. b. $-7x(-6y) = [-7(-6)](x \cdot y)$
$= 42xy$

    c. $4d \cdot 3e \cdot 5 = (4 \cdot 3 \cdot 5)(d \cdot e)$
$= 60de$

    d. $(4s)8 = (4 \cdot 8)s$
$= 32s$

    e. $-1(-e)(2) = [-1(2)](-e)$
$= -2(-e)$
$= 2e$

    f. $7x \cdot 7y = (7 \cdot 7)(x \cdot y)$
$= 49xy$

    g. $4 \cdot 3k \cdot 7 = (4 \cdot 3 \cdot 7)k$
$= 84k$

    h. $(-10t)(-10) = [-10(-10)]t$
$= 100t$

17. a. $-(6t - 4) = -1(6t - 4)$
$= -1(6t) - (-1)(4)$
$= -6t - (-4)$
$= -6t + 4$

    b. $-(5 + x) = -1(5 + x)$
$= -1(5) + (-1)(x)$
$= -5 + (-x)$
$= -5 - x$

    c. $-(6t - 3s + 1) = -1(6t - 3s + 1)$
$= -1(6t) - (-1)(3s) + (-1)(1)$
$= -6t - (-3s) + (-1)$
$= -6t + 3s - 1$

17. d. $-(-5a-3) = -1(-5a-3)$
$$= -1(-5a) - (-1)(3)$$
$$= 5a - (-3)$$
$$= 5a + 3$$

19. a. In $5x - 6y^2$, x is used as a factor.

  b. In $2b - x + 6$, x is used as a term.

  c. In $6xy$, x is used as a factor.

  d. In $-36 + x + b$, x is used as a term.

21. a. $3x + 4x = 7x$

  b. $6r - 9r = -3r$

  c. $-3t - 6t = -9t$

  d. $2z + (-5z) = -3z$

  e. $6x - x = 5x$

  f. $-6y - 7y - (-y) = -12y$

  g. $5w - 8 - 4w + 3 = w - 5$

  h. $-5x - 5y - x + 7y = -6x + 2y$

23. a. $7(y+6) + 3(2y+2) = 7y + 42 + 6y + 6$
$$= 13y + 48$$

  b. $-4(t-7) - (t+6) = -4t + 28 - t - 6$
$$= -5t + 22$$

  c. $5x - 2(x-6) = 5x - 2x + 12$
$$= 3x + 12$$

  d. $6f + 7(12 - 8f) = 6f + 84 - 56f$
$$= -50f + 84$$

25.
$$-4x+6=2(x+12)$$
$$-4(-3)+6\overset{?}{=}2(-3+12)$$
$$12+6\overset{?}{=}2(9)$$
$$18=18$$
Yes, −3 is a solution.

27. Steps of the strategy for solving equations:

1. Use the distributive property to remove any parentheses.

2. Combine like terms on either side of the equation.

3. Apply the addition or subtraction properties of equality to get the variables on one side of the = sign and the constants on the other.

4. Continue to combine like terms when necessary.

5. Undo the operations of multiplication and division to isolate the variable.

29. Let x = the number of miles she biked.
$x-8$ = the number of miles she jogged.
$$x+x-8=18$$
$$2x-8=18$$
$$2x-8+8=18+8$$
$$2x=26$$
$$\frac{2x}{2}=\frac{26}{2}$$
$$x=13$$
She jogged 5 miles and biked 13 miles.

31. Let x = attendance the first day
$2x$ = attendance the second day
$3x$ = attendance the third day
$$x+2x+3x=6{,}600$$
$$6x=6{,}600$$
$$x=1{,}100$$
The attendance was 1,100 on the first day, 2,200 on the second day, and 3,300 on the third day.

33. a. $(4h)^3 = 4h \cdot 4h \cdot 4h$

b. $5 \cdot 5 \cdot d \cdot d \cdot d \cdot m \cdot m \cdot m \cdot m = 5^2 d^3 m^4$

35. a. $2b^2 \cdot 4b^5 = (2 \cdot 4)(b^2 \cdot b^5)$
$$= 8b^{2+5}$$
$$= 8b^7$$

b. $-6x^3(4x) = (-6 \cdot 4)(x^3 \cdot x)$
$$= -24x^{3+1}$$
$$= -24x^4$$

35. c. $-2f^2(-4f)(3f^4) = [-2(-4)(3)](f^2 \bullet f \bullet f^4)$

$$= 24f^{2+1+4}$$
$$= 24f^7$$

d. $-ab \bullet b \bullet a = -1(a \bullet a)(b \bullet b)$

$$= -1 \bullet a^{1+1}b^{1+1}$$
$$= -a^2b^2$$

e. $xy^4 \bullet xy^2 = (x \bullet x)(y^4 \bullet y^2)$

$$= x^{1+1}y^{4+2}$$
$$= x^2y^6$$

f. $(mn)(mn) = (m \bullet m)(n \bullet n)$

$$= m^{1+1}n^{1+1}$$
$$= m^2n^2$$

g. $3z^3 \bullet 9m^3z^4 = (3 \bullet 9)(m^3)(z^3 \bullet z^4)$

$$= 27m^3z^{3+4}$$
$$= 27m^3z^7$$

h. $-5c\,d(4c^2d^5) = (-5 \bullet 4)(c \bullet c^2)(d \bullet d^5)$

$$= -20c^{1+2}d^{1+5}$$
$$= -20c^3d^6$$

37. a. $\left(c^4\right)^5\left(c^2\right)^3 = c^{4 \bullet 5}c^{2 \bullet 3}$

$$= c^{20}c^6$$
$$= c^{20+6}$$
$$= c^{26}$$

b. $\left(3s^2\right)^3\left(2s^3\right)^2 = 3^3s^{2 \bullet 3} \bullet 2^2s^{3 \bullet 2}$

$$= 27s^6 \bullet 4s^6$$
$$= (27 \bullet 4)(s^6 \bullet s^6)$$
$$= 108s^{6+6}$$
$$= 108s^{12}$$

37. c. $\left(c^4c^3\right)^2 = \left(c^{4+3}\right)^2$

$$= \left(c^7\right)^2$$
$$= c^{7 \bullet 2}$$
$$= c^{14}$$

d. $\left(2xx^2\right)^3 = \left(2x^{1+2}\right)^3$

$$= \left(2x^3\right)^3$$
$$= 2^3x^{3 \bullet 3}$$
$$= 8x^9$$

KEY CONCEPT Order of Operations

The rules for the order of operations are necessary because <u>we can obtain different answers for the same problem</u>.

1. Evaluate all <u>powers</u>.

2. Do all <u>multiplications</u> and divisions as they occur from <u>left</u> to right.

3. Do all additions and <u>subtractions</u> as they occur from left to <u>right</u>.

1. addition, subtraction, power

3. multiplication, subtraction, division

5.
$$-10 + 4 - 3^2 = -10 + 4 - 9$$
$$= -6 - 9$$
$$= -15$$

7.
$$-2(-3) - 12 \div 6 \cdot 3 = 6 - 2 \cdot 3$$
$$= 6 - 6$$
$$= 0$$

9.
$$2(4 + 3 \cdot 2)^2 - (-6) = 2(4 + 6)^2 - (-6)$$
$$= 2(10)^2 - (-6)$$
$$= 2(100) - (-6)$$
$$= 200 - (-6)$$
$$= 200 + 6$$
$$= 206$$

CHAPTER 3 TEST

1. a. $r - 2$

b. $3xy$

3. a.
$$\frac{x - 16}{x} = \frac{4 - 16}{4}$$
$$= \frac{-12}{4}$$
$$= -3$$

3. b.
$$2t^2 - 3(t - s) = 2(-2)^2 - 3(-2 - 4)$$
$$= 2(4) - 3(-6)$$
$$= 8 - (-18)$$
$$= 8 + [-(-18)]$$
$$= 8 + 18$$
$$= 26$$

5. $p = r - c$
$$p = (40,000 + 15,000) - (13,000 + 5,000)$$
$$= 55,000 - 18,000$$
$$= 37,000$$
The profit was \$37,000.

7. $\text{mean} = \dfrac{4 + (-3) + (-5) + 1 + (-2)}{5}$
$$= \frac{-5}{5}$$
$$= -1$$
The average reading was –1.

9. $C = \dfrac{5(F - 32)}{9}$
$$= \frac{5(59 - 32)}{9}$$
$$= \frac{5(27)}{9}$$
$$= \frac{135}{9}$$
$$= 15$$
The temperature was 15° C.

11. a. In $5xy$, x is used as a factor.

b. In $8y + x + 6$, x is used as a term.

13. In $8x^2 - 4x - 6$, the terms are $8x^2$, $-4x$, and -6.

15. $4(y + 3) - 5(2y + 3) = 4y + 12 - 10y - 15$
$$= -6y - 3$$

17.
$$2(4x-1)=3(4-2x)$$
$$8x-2=12-6x$$
$$8x-2+6x=12-6x+6x$$
$$14x-2=12$$
$$14x-2+2=12+2$$
$$14x=14$$
$$\frac{14x}{14}=\frac{14}{14}$$
$$x=1$$

19. Let $x =$ the length of each class session.
$$4x+2=14$$
$$4x+2-2=14-2$$
$$4x=12$$
$$\frac{4x}{4}=\frac{12}{4}$$
$$x=3$$
Each class session is 3 hours long.

21. a. $h^2 h^4 = h^{2+4}$
$$= h^6$$

b. $-7x^3(4x^2) = (-7 \bullet 4)(x^3 \bullet x^2)$
$$= -28x^{3+2}$$
$$= -28x^5$$

c. $b^2 \bullet b \bullet b^5 = b^{2+1+5}$
$$= b^8$$

d. $-3g^2 k^3(-8g^3 k^{10}) = (-3 \bullet -8)(g^2 \bullet g^3)(k^3 \bullet k^{10})$
$$= 24g^{2+3}k^{3+10}$$
$$= 24g^5 k^{13}$$

STUDY SET 4.1

VOCABULARY

1. For the fraction $\frac{7}{8}$, 7 is the <u>numerator</u> and 8 is the <u>denominator</u>.

3. We can simplify a fraction that is not in lowest terms by <u>dividing</u> the numerator and the denominator by the <u>same</u> number.

5. Two fractions are <u>equivalent</u> if they have the same value.

7. Multiplying the numerator and denominator of a fraction by a number to obtain an equivalent fraction that involves larger numbers or more complex terms is called expressing the fraction in <u>higher terms.</u>

CONCEPTS

9. They indicate that the 5's were divided by 5 and that each result is 1.

11. The illustration shows equivalent fractions: $\dfrac{2}{6} = \dfrac{1}{3}$

13. a. The first divides out a common factor that we know divides into both; the second uses the prime factorizations and divides out all common factors.

b. Yes, the results are the same.

15. The 2's in the numerator and denominator aren't common factors.

17. $\dfrac{c}{d} = \dfrac{c \cdot t}{d \cdot t}$ and $\dfrac{c}{d} = \dfrac{c \div t}{d \div t}$

19. $\dfrac{18}{24} = \dfrac{3 \cdot 3 \cdot 2}{3 \cdot 2 \cdot 2 \cdot 2}$

$= \dfrac{\overset{1}{\cancel{3}} \cdot 3 \cdot \overset{1}{\cancel{2}}}{\cancel{3} \cdot \cancel{2} \cdot 2 \cdot 2}$

$= \dfrac{3}{4}$

PRACTICE

21. $\dfrac{3}{9} = \dfrac{3 \cdot 1}{3 \cdot 3}$

$= \dfrac{\overset{1}{\cancel{3}} \cdot 1}{\underset{1}{\cancel{3}} \cdot 3}$

$= \dfrac{1}{3}$

23. $-\dfrac{7}{21} = -\dfrac{7 \cdot 1}{7 \cdot 3}$

$= -\dfrac{\overset{1}{\cancel{7}} \cdot 1}{\underset{1}{\cancel{7}} \cdot 3}$

$= -\dfrac{1}{3}$

25. $\dfrac{20}{30} = \dfrac{10 \cdot 2}{10 \cdot 3}$

$= \dfrac{\overset{1}{\cancel{10}} \cdot 2}{\underset{1}{\cancel{10}} \cdot 3}$

$= \dfrac{2}{3}$

27. $\dfrac{15}{6} = \dfrac{5 \cdot 3}{3 \cdot 2}$

$\qquad = \dfrac{5 \cdot \overset{1}{\cancel{3}}}{\underset{1}{\cancel{3}} \cdot 2}$

$\qquad = \dfrac{5}{2}$

29. $-\dfrac{28}{56} = -\dfrac{7 \cdot 2 \cdot 2}{7 \cdot 2 \cdot 2 \cdot 2}$

$\qquad = -\dfrac{\overset{1}{\cancel{7}} \cdot \overset{1}{\cancel{2}} \cdot \overset{1}{\cancel{2}}}{\underset{1}{\cancel{7}} \cdot \underset{1}{\cancel{2}} \cdot \underset{1}{\cancel{2}} \cdot 2}$

$\qquad = -\dfrac{1}{2}$

31. $\dfrac{90}{105} = \dfrac{5 \cdot 3 \cdot 3 \cdot 2}{7 \cdot 5 \cdot 3}$

$\qquad = \dfrac{\overset{1}{\cancel{5}} \cdot \overset{1}{\cancel{3}} \cdot 3 \cdot 2}{7 \cdot \underset{1}{\cancel{5}} \cdot \underset{1}{\cancel{3}}}$

$\qquad = \dfrac{6}{7}$

33. $\dfrac{60}{108} = \dfrac{5 \cdot 3 \cdot 2 \cdot 2}{3 \cdot 3 \cdot 3 \cdot 2 \cdot 2}$

$\qquad = \dfrac{5 \cdot \overset{1}{\cancel{3}} \cdot \overset{1}{\cancel{2}} \cdot \overset{1}{\cancel{2}}}{\underset{1}{\cancel{3}} \cdot 3 \cdot 3 \cdot \underset{1}{\cancel{2}} \cdot \underset{1}{\cancel{2}}}$

$\qquad = \dfrac{5}{9}$

35. $\dfrac{180}{210} = \dfrac{30 \cdot 6}{30 \cdot 7}$

$\qquad = \dfrac{\overset{1}{\cancel{30}} \cdot 6}{\underset{1}{\cancel{30}} \cdot 7}$

$\qquad = \dfrac{6}{7}$

37. $\dfrac{55}{67}$ in lowest terms

39. $\dfrac{36}{96} = \dfrac{12 \cdot 3}{12 \cdot 8}$

$\qquad = \dfrac{\overset{1}{\cancel{12}} \cdot 3}{\underset{1}{\cancel{12}} \cdot 8}$

$\qquad = \dfrac{3}{8}$

41. $\dfrac{25x^2}{35x} = \dfrac{5 \cdot 5 \cdot x \cdot x}{7 \cdot 5 \cdot x}$

$\qquad = \dfrac{\overset{1}{\cancel{5}} \cdot 5 \cdot x \cdot \overset{1}{\cancel{x}}}{7 \cdot \underset{1}{\cancel{5}} \cdot \underset{1}{\cancel{x}}}$

$\qquad = \dfrac{5x}{7}$

43. $\dfrac{12t}{15t} = \dfrac{4 \cdot 3 \cdot t}{5 \cdot 3 \cdot t}$

$\qquad = \dfrac{4 \cdot \overset{1}{\cancel{3}} \cdot \overset{1}{\cancel{t}}}{5 \cdot \underset{1}{\cancel{3}} \cdot \underset{1}{\cancel{t}}}$

$\qquad = \dfrac{4}{5}$

45. $\dfrac{6a}{7a} = \dfrac{6 \cdot a}{7 \cdot a}$

$\qquad = \dfrac{6 \cdot \overset{1}{\cancel{a}}}{7 \cdot \underset{1}{\cancel{a}}}$

$\qquad = \dfrac{6}{7}$

47. $\dfrac{7xy}{8xy} = \dfrac{7 \cdot x \cdot y}{8 \cdot x \cdot y}$

$= \dfrac{7 \cdot \overset{1}{\cancel{x}} \cdot \overset{1}{\cancel{y}}}{8 \cdot \underset{1}{\cancel{x}} \cdot \underset{1}{\cancel{y}}}$

$= \dfrac{7}{8}$

49. $\dfrac{10rs}{30} = \dfrac{10 \cdot r \cdot s}{10 \cdot 3}$

$= \dfrac{\overset{1}{\cancel{10}} \cdot r \cdot s}{\underset{1}{\cancel{10}} \cdot 3}$

$= \dfrac{rs}{3}$

51. $\dfrac{15st^3}{25xt^3} = \dfrac{5 \cdot 3 \cdot s \cdot t \cdot t \cdot t}{5 \cdot 5 \cdot x \cdot t \cdot t \cdot t}$

$= \dfrac{\overset{1}{\cancel{5}} \cdot 3 \cdot s \cdot \overset{1}{\cancel{t}} \cdot \overset{1}{\cancel{t}} \cdot \overset{1}{\cancel{t}}}{\underset{1}{\cancel{5}} \cdot 5 \cdot x \cdot \underset{1}{\cancel{t}} \cdot \underset{1}{\cancel{t}} \cdot \underset{1}{\cancel{t}}}$

$= \dfrac{3s}{5x}$

53. $\dfrac{35r^2t}{28rt^2} = \dfrac{7 \cdot 5 \cdot r \cdot r \cdot t}{7 \cdot 4 \cdot r \cdot t \cdot t}$

$= \dfrac{\overset{1}{\cancel{7}} \cdot 5 \cdot \overset{1}{\cancel{r}} \cdot r \cdot \overset{1}{\cancel{t}}}{\underset{1}{\cancel{7}} \cdot 4 \cdot \underset{1}{\cancel{r}} \cdot \underset{1}{\cancel{t}} \cdot t}$

$= \dfrac{5r}{4t}$

55. $\dfrac{56p^4}{28p^6} = \dfrac{28 \cdot 2 \cdot p \cdot p \cdot p \cdot p}{28 \cdot p \cdot p \cdot p \cdot p \cdot p \cdot p}$

$= \dfrac{\overset{1}{\cancel{28}} \cdot 2 \cdot \overset{1}{\cancel{p}} \cdot \overset{1}{\cancel{p}} \cdot \overset{1}{\cancel{p}} \cdot \overset{1}{\cancel{p}}}{\underset{1}{\cancel{28}} \cdot \underset{1}{\cancel{p}} \cdot \underset{1}{\cancel{p}} \cdot \underset{1}{\cancel{p}} \cdot \underset{1}{\cancel{p}} \cdot p \cdot p}$

$= \dfrac{2}{p^2}$

57. $\dfrac{7}{8} = \dfrac{7 \cdot 5}{8 \cdot 5}$

$= \dfrac{35}{40}$

59. $\dfrac{4}{5} = \dfrac{4 \cdot 7}{5 \cdot 7}$

$= \dfrac{28}{35}$

61. $\dfrac{-5}{6} = \dfrac{-5 \cdot 9}{6 \cdot 9}$

$= -\dfrac{45}{54}$

63. $-\dfrac{1}{2} = -\dfrac{1 \cdot 15}{2 \cdot 15}$

$= -\dfrac{15}{30}$

65. $\dfrac{2}{7} = \dfrac{2 \cdot 2x}{7 \cdot 2x}$

$= \dfrac{4x}{14x}$

67. $\dfrac{9}{10} = \dfrac{9 \cdot 6t}{10 \cdot 6t}$

$= \dfrac{54t}{60t}$

69.
$$-\frac{5}{4s} = -\frac{5 \cdot 5}{4s \cdot 5}$$
$$= -\frac{25}{20s}$$

75.
$$\frac{-6}{1} = -\frac{6 \cdot 8}{1 \cdot 8}$$
$$= -\frac{48}{8}$$

71.
$$\frac{-2}{15} = -\frac{2 \cdot 3y}{15 \cdot 3y}$$
$$= -\frac{6y}{45y}$$

77.
$$\frac{4a}{1} = \frac{4a \cdot 9}{1 \cdot 9}$$
$$= \frac{36a}{9}$$

73.
$$\frac{3}{1} = \frac{3 \cdot 5}{1 \cdot 5}$$
$$= \frac{15}{5}$$

79.
$$\frac{-2t}{1} = -\frac{2t \cdot 2}{1 \cdot 2}$$
$$= -\frac{4t}{2}$$

APPLICATIONS

81. $\dfrac{3}{5}$

83. $\dfrac{3}{4}$

85.

Name	Total time to complete the job alone	Time worked alone	Amount of job completed
Bob	10 hours	7 hours	$\dfrac{7}{10}$
Ali	8 hours	1 hour	$\dfrac{1}{8}$

87. The diagram gives the dates of the full, half, and new moons.

89. To get from A to B would take one quarter turn to the left or three-quarters of a turn to the right.

91.

Snacks
Potato chips
Peanuts
Pretzels
Tortilla chips

REVIEW

97.
$$-5x+1=16$$
$$-5x+1-1=16-1$$
$$-5x=15$$
$$\frac{-5x}{-5}=\frac{15}{-5}$$
$$x=-3$$

99. 564,000

101. d dimes have a value of $10d¢$.

103. $(-10)^2=(-10)(-10)$
$$=100$$

STUDY SET Section 4.2
VOCABULARY

1. The word *of* in mathematics usually means <u>multiply</u>.

3. The result of a multiplication problem is called the <u>product</u>.

5. In a triangle, b stands for the length of the <u>base</u> and h stands for the <u>height</u>.

CONCEPTS

7. $\dfrac{a}{b}\cdot\dfrac{c}{d}=\dfrac{ac}{bd}$

9. a.

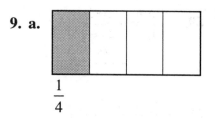

$\dfrac{1}{4}$

b.

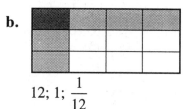

$12;\ 1;\ \dfrac{1}{12}$

11. If we evaluated $\left(-\dfrac{4}{5}\right)^{10}$ the result would be positive.

13. a. $\dfrac{1}{2}x=\dfrac{x}{2}$ is true

b. $\dfrac{2t}{3}=\dfrac{2}{3}t$ is true

c. $-\dfrac{3}{8}a=-\dfrac{3}{8a}$ is false

d. $\dfrac{-4e}{7}=-\dfrac{4e}{7}$ is true

NOTATION

15. $\dfrac{5}{8}\cdot\dfrac{7}{15}=\dfrac{5\cdot7}{8\cdot15}$
$$=\dfrac{5\cdot7}{8\cdot5\cdot3}$$
$$=\dfrac{\overset{1}{\cancel{5}}\cdot7}{8\cdot\underset{1}{\cancel{5}}\cdot3}$$
$$=\dfrac{7}{24}$$

PRACTICE

17. $\dfrac{1}{4} \cdot \dfrac{1}{2} = \dfrac{1 \cdot 1}{4 \cdot 2}$

$\qquad = \dfrac{1}{8}$

19. $\dfrac{3}{8} \cdot \dfrac{7}{16} = \dfrac{3 \cdot 7}{8 \cdot 16}$

$\qquad = \dfrac{21}{128}$

21. $\dfrac{2}{3} \cdot \dfrac{6}{7} = \dfrac{2 \cdot 6}{3 \cdot 7}$

$\qquad = \dfrac{2 \cdot 3 \cdot 2}{3 \cdot 7}$

$\qquad = \dfrac{2 \cdot \overset{1}{\cancel{3}} \cdot 2}{\underset{1}{\cancel{3}} \cdot 7}$

$\qquad = \dfrac{4}{7}$

23. $\dfrac{14}{15} \cdot \dfrac{11}{8} = \dfrac{14 \cdot 11}{15 \cdot 8}$

$\qquad = \dfrac{7 \cdot 2 \cdot 11}{15 \cdot 4 \cdot 2}$

$\qquad = \dfrac{7 \cdot \overset{1}{\cancel{2}} \cdot 11}{15 \cdot 4 \cdot \underset{1}{\cancel{2}}}$

$\qquad = \dfrac{77}{60}$

25. $-\dfrac{15}{24} \cdot \dfrac{8}{25} = -\dfrac{15 \cdot 8}{24 \cdot 25}$

$\qquad = -\dfrac{5 \cdot 3 \cdot 8}{8 \cdot 3 \cdot 5 \cdot 5}$

$\qquad = -\dfrac{\overset{1}{\cancel{5}} \cdot \overset{1}{\cancel{3}} \cdot \overset{1}{\cancel{8}}}{\underset{1}{\cancel{8}} \cdot \underset{1}{\cancel{3}} \cdot \underset{1}{\cancel{5}} \cdot 5}$

$\qquad = -\dfrac{1}{5}$

27. $\left(-\dfrac{11}{21}\right)\left(-\dfrac{14}{33}\right) = \dfrac{11 \cdot 14}{21 \cdot 33}$

$\qquad = \dfrac{11 \cdot 7 \cdot 2}{7 \cdot 3 \cdot 11 \cdot 3}$

$\qquad = \dfrac{\overset{1}{\cancel{11}} \cdot \overset{1}{\cancel{7}} \cdot 2}{\underset{1}{\cancel{7}} \cdot 3 \cdot \underset{1}{\cancel{11}} \cdot 3}$

$\qquad = \dfrac{2}{9}$

29. $\dfrac{7}{10}\left(\dfrac{20}{21}\right) = \dfrac{7 \cdot 20}{10 \cdot 21}$

$\qquad = \dfrac{7 \cdot 10 \cdot 2}{10 \cdot 7 \cdot 3}$

$\qquad = \dfrac{\overset{1}{\cancel{7}} \cdot \overset{1}{\cancel{10}} \cdot 2}{\underset{1}{\cancel{10}} \cdot \underset{1}{\cancel{7}} \cdot 3}$

$\qquad = \dfrac{2}{3}$

31. $\dfrac{3}{4} \cdot \dfrac{4}{3} = \dfrac{3 \cdot 4}{4 \cdot 3}$

$\qquad = \dfrac{\overset{1}{\cancel{3}} \cdot \overset{1}{\cancel{4}}}{\underset{1}{\cancel{4}} \cdot \underset{1}{\cancel{3}}}$

$\qquad = \dfrac{1}{1}$

$\qquad = 1$

33. $\dfrac{1}{3} \cdot \dfrac{15}{16} \cdot \dfrac{4}{25} = \dfrac{1 \cdot 15 \cdot 4}{3 \cdot 16 \cdot 25}$

$\qquad = \dfrac{1 \cdot 5 \cdot 3 \cdot 4}{3 \cdot 4 \cdot 4 \cdot 5 \cdot 5}$

$\qquad = \dfrac{1 \cdot \overset{1}{\cancel{5}} \cdot \overset{1}{\cancel{3}} \cdot \overset{1}{\cancel{4}}}{\underset{1}{\cancel{3}} \cdot 4 \cdot 4 \cdot \underset{1}{\cancel{5}} \cdot 5}$

$\qquad = \dfrac{1}{20}$

35. $\left(\dfrac{2}{3}\right)\left(-\dfrac{1}{16}\right)\left(-\dfrac{4}{5}\right) = \dfrac{2 \cdot 1 \cdot 4}{3 \cdot 16 \cdot 5}$

$\qquad = \dfrac{2 \cdot 1 \cdot 4}{3 \cdot 4 \cdot 2 \cdot 2 \cdot 5}$

$\qquad = \dfrac{\overset{1}{\cancel{2}} \cdot 1 \cdot \overset{1}{\cancel{4}}}{3 \cdot \cancel{4} \cdot \underset{1}{\cancel{2}} \cdot 2 \cdot 5}$

$\qquad = \dfrac{1}{30}$

37. $\dfrac{5}{6} \cdot 18 = \dfrac{5}{6} \cdot \dfrac{18}{1}$

$\qquad = \dfrac{5 \cdot 18}{6 \cdot 1}$

$\qquad = \dfrac{5 \cdot 6 \cdot 3}{6 \cdot 1}$

$\qquad = \dfrac{5 \cdot \overset{1}{\cancel{6}} \cdot 3}{\underset{1}{\cancel{6}} \cdot 1}$

$\qquad = \dfrac{15}{1}$

$\qquad = 15$

39. $15\left(-\dfrac{4}{5}\right) = \dfrac{15}{1}\left(-\dfrac{4}{5}\right)$

$\qquad = -\dfrac{15 \cdot 4}{1 \cdot 5}$

$\qquad = -\dfrac{5 \cdot 3 \cdot 4}{1 \cdot 5}$

$\qquad = -\dfrac{\overset{1}{\cancel{5}} \cdot 3 \cdot 4}{1 \cdot \underset{1}{\cancel{5}}}$

$\qquad = -\dfrac{12}{1}$

$\qquad = -12$

41. $\dfrac{5x}{12} \cdot \dfrac{1}{6x} = \dfrac{5x \cdot 1}{12 \cdot 6x}$

$\qquad = \dfrac{5 \cdot x \cdot 1}{12 \cdot 6 \cdot x}$

$\qquad = \dfrac{5 \cdot \overset{1}{\cancel{x}} \cdot 1}{12 \cdot 6 \cdot \underset{1}{\cancel{x}}}$

$\qquad = \dfrac{5}{72}$

43. $\dfrac{b}{12} \cdot \dfrac{3}{10b} = \dfrac{b \cdot 3}{12 \cdot 10b}$

$\qquad = \dfrac{b \cdot 3}{4 \cdot 3 \cdot 10 \cdot b}$

$\qquad = \dfrac{\overset{1}{\cancel{b}} \cdot \overset{1}{\cancel{3}}}{4 \cdot \underset{1}{\cancel{3}} \cdot 10 \cdot \underset{1}{\cancel{b}}}$

$\qquad = \dfrac{1}{40}$

45. $\dfrac{1}{3} \cdot 3d = \dfrac{1}{3} \cdot \dfrac{3d}{1}$

$\qquad = \dfrac{1 \cdot 3d}{3 \cdot 1}$

$\qquad = \dfrac{1 \cdot 3 \cdot d}{3 \cdot 1}$

$\qquad = \dfrac{1 \cdot \overset{1}{\cancel{3}} \cdot d}{\underset{1}{\cancel{3}} \cdot 1}$

$\qquad = \dfrac{d}{1}$

$\qquad = d$

47. $\dfrac{2}{3} \cdot \dfrac{3s}{2} = \dfrac{2 \cdot 3s}{3 \cdot 2}$

$\qquad = \dfrac{2 \cdot 3 \cdot s}{3 \cdot 2}$

$\qquad = \dfrac{\overset{1}{\cancel{2}} \cdot \overset{1}{\cancel{3}} \cdot s}{\underset{1}{\cancel{3}} \cdot \underset{1}{\cancel{2}}}$

$\qquad = \dfrac{s}{1}$

$\qquad = s$

49. $-\dfrac{5}{6} \cdot \dfrac{6}{5}c = -\dfrac{5}{6} \cdot \dfrac{6}{5} \cdot \dfrac{c}{1}$

$\qquad = -\dfrac{5 \cdot 6 \cdot c}{6 \cdot 5 \cdot 1}$

$\qquad = -\dfrac{\overset{1}{\cancel{5}} \cdot \overset{1}{\cancel{6}} \cdot c}{\underset{1}{\cancel{6}} \cdot \underset{1}{\cancel{5}} \cdot 1}$

$\qquad = -\dfrac{c}{1}$

$\qquad = -c$

51. $\dfrac{xy}{2} \cdot \dfrac{4}{9x} = \dfrac{xy \cdot 4}{2 \cdot 9x}$

$\qquad = \dfrac{x \cdot y \cdot 2 \cdot 2}{2 \cdot 9 \cdot x}$

$\qquad = \dfrac{\overset{1}{\cancel{x}} \cdot y \cdot \overset{1}{\cancel{2}} \cdot 2}{\underset{1}{\cancel{2}} \cdot 9 \cdot \underset{1}{\cancel{x}}}$

$\qquad = \dfrac{2y}{9}$

53. $4ef \cdot \dfrac{e}{2f} = \dfrac{4ef}{1} \cdot \dfrac{e}{2f}$

$\qquad = \dfrac{4ef \cdot e}{1 \cdot 2f}$

$\qquad = \dfrac{2 \cdot 2 \cdot e \cdot f \cdot e}{1 \cdot 2 \cdot f}$

$\qquad = \dfrac{\overset{1}{\cancel{2}} \cdot 2 \cdot e \cdot \overset{1}{\cancel{f}} \cdot e}{1 \cdot \underset{1}{\cancel{2}} \cdot \underset{1}{\cancel{f}}}$

$\qquad = \dfrac{2e^2}{1}$

$\qquad = 2e^2$

55. $\dfrac{5y^2}{8x}\left(\dfrac{2x^3}{15y}\right) = \dfrac{5y^2 \cdot 2x^3}{8x \cdot 15y}$

$\qquad = \dfrac{5 \cdot y \cdot y \cdot 2 \cdot x \cdot x \cdot x}{4 \cdot 2 \cdot x \cdot 5 \cdot 3 \cdot y}$

$\qquad = \dfrac{\overset{1}{\cancel{5}} \cdot \overset{1}{\cancel{y}} \cdot y \cdot \overset{1}{\cancel{2}} \cdot \overset{1}{\cancel{x}} \cdot x \cdot x}{4 \cdot \underset{1}{\cancel{2}} \cdot \underset{1}{\cancel{x}} \cdot \underset{1}{\cancel{5}} \cdot 3 \cdot \underset{1}{\cancel{y}}}$

$\qquad = \dfrac{x^2 y}{12}$

57.
$$-\frac{5c}{6cd^2} \cdot \frac{12d^4}{c} = -\frac{5c \cdot 12d^4}{6cd^2 \cdot c}$$
$$= -\frac{5 \cdot c \cdot 6 \cdot 2 \cdot d \cdot d \cdot d \cdot d}{6 \cdot c \cdot d \cdot d \cdot c}$$
$$= -\frac{5 \cdot \overset{1}{\cancel{c}} \cdot \overset{1}{\cancel{6}} \cdot 2 \cdot \overset{1}{\cancel{d}} \cdot \overset{1}{\cancel{d}} \cdot d \cdot d}{\underset{1}{\cancel{6}} \cdot \underset{1}{\cancel{c}} \cdot \underset{1}{\cancel{d}} \cdot \underset{1}{\cancel{d}} \cdot c}$$
$$= -\frac{10d^2}{c}$$

59.
$$-\frac{4h^2}{5}\left(-\frac{15s}{16h^3}\right) = \frac{4h^2 \cdot 15s}{5 \cdot 16h^3}$$
$$= \frac{4 \cdot h \cdot h \cdot 5 \cdot 3 \cdot s}{5 \cdot 4 \cdot 4 \cdot h \cdot h \cdot h}$$
$$= \frac{\overset{1}{\cancel{4}} \cdot \overset{1}{\cancel{h}} \cdot \overset{1}{\cancel{h}} \cdot \overset{1}{\cancel{5}} \cdot 3 \cdot s}{\underset{1}{\cancel{5}} \cdot \underset{1}{\cancel{4}} \cdot 4 \cdot \underset{1}{\cancel{h}} \cdot \underset{1}{\cancel{h}} \cdot h}$$
$$= \frac{3s}{4h}$$

61.
$$\frac{5}{6} \cdot x = \frac{5}{6} \cdot \frac{x}{1}$$
$$= \frac{5x}{6}$$
$$\frac{5x}{6} \text{ or } \frac{5}{6}x$$

63.
$$-\frac{8}{9} \cdot v = -\frac{8}{9} \cdot \frac{v}{1}$$
$$= -\frac{8v}{9}$$
$$-\frac{8v}{9} \text{ or } -\frac{8}{9}v$$

65.
$$\left(\frac{2}{3}\right)^2 = \left(\frac{2}{3}\right)\left(\frac{2}{3}\right)$$
$$= \frac{2 \cdot 2}{3 \cdot 3}$$
$$= \frac{4}{9}$$

67.
$$\left(-\frac{5}{9}\right)^2 = \left(-\frac{5}{9}\right)\left(-\frac{5}{9}\right)$$
$$= \frac{5 \cdot 5}{9 \cdot 9}$$
$$= \frac{25}{81}$$

69. $\left(\dfrac{4m}{3}\right)^2 = \left(\dfrac{4m}{3}\right)\left(\dfrac{4m}{3}\right)$

$= \dfrac{4m \cdot 4m}{3 \cdot 3}$

$= \dfrac{16m^2}{9}$

71. $\left(-\dfrac{3r}{4}\right)^3 = \left(-\dfrac{3r}{4}\right)\left(-\dfrac{3r}{4}\right)\left(-\dfrac{3r}{4}\right)$

$= -\dfrac{3r \cdot 3r \cdot 3r}{4 \cdot 4 \cdot 4}$

$= -\dfrac{27r^3}{64}$

73.

	$\dfrac{1}{2}$	$\dfrac{1}{3}$	$\dfrac{1}{4}$	$\dfrac{1}{5}$	$\dfrac{1}{6}$
$\dfrac{1}{2}$	$\dfrac{1}{4}$	$\dfrac{1}{6}$	$\dfrac{1}{8}$	$\dfrac{1}{10}$	$\dfrac{1}{12}$
$\dfrac{1}{3}$	$\dfrac{1}{6}$	$\dfrac{1}{9}$	$\dfrac{1}{12}$	$\dfrac{1}{15}$	$\dfrac{1}{18}$
$\dfrac{1}{4}$	$\dfrac{1}{8}$	$\dfrac{1}{12}$	$\dfrac{1}{16}$	$\dfrac{1}{20}$	$\dfrac{1}{24}$
$\dfrac{1}{5}$	$\dfrac{1}{10}$	$\dfrac{1}{15}$	$\dfrac{1}{20}$	$\dfrac{1}{25}$	$\dfrac{1}{30}$
$\dfrac{1}{6}$	$\dfrac{1}{12}$	$\dfrac{1}{18}$	$\dfrac{1}{24}$	$\dfrac{1}{30}$	$\dfrac{1}{36}$

75. $A = \dfrac{1}{2}bh$

$A = \dfrac{1}{2}(10)(3)$

$= \dfrac{1}{2}\left(\dfrac{10}{1}\right)\left(\dfrac{3}{1}\right)$

$= \dfrac{1 \cdot 10 \cdot 3}{2 \cdot 1 \cdot 1}$

$= \dfrac{1 \cdot 5 \cdot 2 \cdot 3}{2 \cdot 1 \cdot 1}$

$= \dfrac{1 \cdot 5 \cdot \overset{1}{2} \cdot 3}{\underset{1}{2} \cdot 1 \cdot 1}$

$= \dfrac{15}{1}$

$A = 15$

The area is 15 ft^2.

77. $A = \dfrac{1}{2}bh$

$A = \dfrac{1}{2}(3)(5)$

$= \dfrac{1}{2}\left(\dfrac{3}{1}\right)\left(\dfrac{5}{1}\right)$

$= \dfrac{1 \cdot 3 \cdot 5}{2 \cdot 1 \cdot 1}$

$A = \dfrac{15}{2}$

The area is $\dfrac{15}{2}$ yd^2.

APPLICATIONS

79. $\dfrac{2}{3}$ of $435 = \dfrac{2}{3} \cdot \dfrac{435}{1}$

$$= \dfrac{2 \cdot 435}{3 \cdot 1}$$

$$= \dfrac{2 \cdot 3 \cdot 145}{3 \cdot 1}$$

$$= \dfrac{2 \cdot \overset{1}{\cancel{3}} \cdot 145}{\underset{1}{\cancel{3}} \cdot 1}$$

$$= \dfrac{290}{1}$$

$$= 290$$

290 votes are needed.

81. $A = s^2 \qquad\qquad A = lw$

$A = \left(\dfrac{7}{8}\right)^2 \qquad A = \dfrac{15}{16} \cdot \dfrac{3}{4}$

$\quad = \dfrac{7}{8} \cdot \dfrac{7}{8} \qquad\quad = \dfrac{15 \cdot 3}{16 \cdot 4}$

$\quad = \dfrac{7 \cdot 7}{8 \cdot 8} \qquad A = \dfrac{45}{64}$

$\quad = \dfrac{49}{64}$

The mallard stamp has the smaller area.

83.

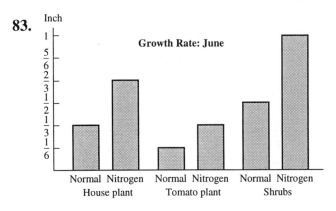

85. $A = \dfrac{1}{2}bh$

$A = \dfrac{1}{2}(7)(12)$

$= \dfrac{1}{2}\left(\dfrac{7}{1}\right)\left(\dfrac{12}{1}\right)$

$= \dfrac{1 \cdot 7 \cdot 12}{2 \cdot 1 \cdot 1}$

$= \dfrac{1 \cdot 7 \cdot 6 \cdot 2}{2 \cdot 1 \cdot 1}$

$= \dfrac{1 \cdot 7 \cdot 6 \cdot \overset{1}{2}}{\underset{1}{2} \cdot 1 \cdot 1}$

$= \dfrac{42}{1}$

$A = 42$

The area of the sail is 42 ft^2.

87. $\dfrac{3}{4} \ of \ 196{,}800{,}000 = \dfrac{3}{4}(196{,}800{,}000)$

$= 147{,}600{,}000$

$147{,}600{,}000$ mi^2 are covered by water.

REVIEW

93. $2(x+7) = 2x + 2(7)$

$\qquad\qquad = 2x + 14$

95. $2x + 6 = 6$

$2(-6) + 6 \overset{?}{=} 6$

$-12 + 6 \overset{?}{=} 6$

$-6 \neq 6$

97. The steps are: analyze the problem, form an equation, solve the equation, state the result, and check the result.

99. $125 = 5^3$

STUDY SET Section 4.3

VOCABULARY

1. Two numbers are called <u>reciprocals</u> if their product is 1.

CONCEPTS

3. $\dfrac{x}{y} \div \dfrac{s}{t} = \dfrac{x}{y} \cdot \underline{\dfrac{t}{s}}$

5.

$$4 \div \frac{1}{3} = 12$$

7.
$$\frac{4}{5} \cdot \frac{5}{4} = \frac{4 \cdot 5}{5 \cdot 4}$$

$$= \frac{\overset{1}{4} \cdot \overset{1}{5}}{\underset{1}{5} \cdot \underset{1}{4}}$$

$$= \frac{1}{1}$$

$$= 1$$

The result is 1.

9. a. $15 \div 3 = 5$

b. $15 \div 3 = 15 \cdot \frac{1}{3}$

$$= \frac{15}{1} \cdot \frac{1}{3}$$

$$= \frac{15 \cdot 1}{1 \cdot 3}$$

$$= \frac{5 \cdot 3 \cdot 1}{1 \cdot 3}$$

$$= \frac{5 \cdot \overset{1}{3} \cdot 1}{\underset{1}{3} \cdot 1}$$

$$= \frac{5}{1}$$

$$= 5$$

c. Division by <u> 3 </u> is the same as multiplication by $\frac{1}{\underline{3}}$.

NOTATION

11.
$$\frac{25}{36} \div \frac{10}{9} = \frac{25}{36} \cdot \frac{9}{10}$$

$$= \frac{25 \cdot 9}{36 \cdot 10}$$

$$= \frac{5 \cdot 5 \cdot 9}{4 \cdot 9 \cdot 2 \cdot 5}$$

$$= \frac{\overset{1}{5} \cdot 5 \cdot \overset{1}{9}}{4 \cdot \underset{1}{9} \cdot 2 \cdot \underset{1}{5}}$$

$$= \frac{5}{8}$$

PRACTICE

13.
$$\frac{1}{2} \div \frac{3}{5} = \frac{1}{2} \cdot \frac{5}{3}$$

$$= \frac{1 \cdot 5}{2 \cdot 3}$$

$$= \frac{5}{6}$$

15.
$$\frac{3}{16} \div \frac{1}{9} = \frac{3}{16} \cdot \frac{9}{1}$$

$$= \frac{3 \cdot 9}{16 \cdot 1}$$

$$= \frac{27}{16}$$

17.
$$\frac{4}{5} \div \frac{4}{5} = \frac{4}{5} \cdot \frac{5}{4}$$

$$= \frac{4 \cdot 5}{5 \cdot 4}$$

$$= \frac{\overset{1}{4} \cdot \overset{1}{5}}{\underset{1}{5} \cdot \underset{1}{4}}$$

$$= \frac{1}{1}$$

$$= 1$$

19. $\left(-\dfrac{7}{4}\right) \div \left(-\dfrac{21}{8}\right) = \left(-\dfrac{7}{4}\right)\left(-\dfrac{8}{21}\right)$

$$= \dfrac{7 \cdot 8}{4 \cdot 21}$$

$$= \dfrac{7 \cdot 4 \cdot 2}{4 \cdot 7 \cdot 3}$$

$$= \dfrac{\overset{1}{7} \cdot \overset{1}{4} \cdot 2}{\underset{1}{4} \cdot \underset{1}{7} \cdot 3}$$

$$= \dfrac{2}{3}$$

21. $3 \div \dfrac{1}{12} = 3 \cdot \dfrac{12}{1}$

$$= \dfrac{3}{1} \cdot \dfrac{12}{1}$$

$$= \dfrac{3 \cdot 12}{1 \cdot 1}$$

$$= \dfrac{36}{1}$$

$$= 36$$

23. $-\dfrac{9}{10} \div \dfrac{4}{15} = -\dfrac{9}{10} \cdot \dfrac{15}{4}$

$$= -\dfrac{9 \cdot 15}{10 \cdot 4}$$

$$= -\dfrac{9 \cdot 5 \cdot 3}{5 \cdot 2 \cdot 4}$$

$$= -\dfrac{9 \cdot \overset{1}{5} \cdot 3}{\underset{1}{5} \cdot 2 \cdot 4}$$

$$= -\dfrac{27}{8}$$

25. $-\dfrac{4}{5} \div (-6) = -\dfrac{4}{5}\left(-\dfrac{1}{6}\right)$

$$= \dfrac{4 \cdot 1}{5 \cdot 6}$$

$$= \dfrac{2 \cdot 2 \cdot 1}{5 \cdot 3 \cdot 2}$$

$$= \dfrac{2 \cdot \overset{1}{2} \cdot 1}{5 \cdot 3 \cdot \underset{1}{2}}$$

$$= \dfrac{2}{15}$$

27. $\dfrac{9}{10} \div \left(-\dfrac{3}{25}\right) = \dfrac{9}{10}\left(-\dfrac{25}{3}\right)$

$$= -\dfrac{9 \cdot 25}{10 \cdot 3}$$

$$= -\dfrac{3 \cdot 3 \cdot 5 \cdot 5}{5 \cdot 2 \cdot 3}$$

$$= -\dfrac{3 \cdot \overset{1}{3} \cdot \overset{1}{5} \cdot 5}{\underset{1}{5} \cdot 2 \cdot \underset{1}{3}}$$

$$= -\dfrac{15}{2}$$

29. $\dfrac{4a}{5} \div \dfrac{3}{2} = \dfrac{4a}{5} \cdot \dfrac{2}{3}$

$$= \dfrac{4a \cdot 2}{5 \cdot 3}$$

$$= \dfrac{8a}{15}$$

31. $\dfrac{t}{8} \div \dfrac{3}{4} = \dfrac{t}{8} \cdot \dfrac{4}{3}$

$$= \dfrac{t \cdot 4}{8 \cdot 3}$$

$$= \dfrac{t \cdot 4}{4 \cdot 2 \cdot 3}$$

$$= \dfrac{t \cdot \overset{1}{4}}{\underset{1}{4} \cdot 2 \cdot 3}$$

$$= \dfrac{t}{6}$$

33. $\dfrac{13}{16b} \div \dfrac{1}{2} = \dfrac{13}{16b} \cdot \dfrac{2}{1}$

$$= \dfrac{13 \cdot 2}{16b \cdot 1}$$

$$= \dfrac{13 \cdot 2}{8 \cdot 2 \cdot b \cdot 1}$$

$$= \dfrac{13 \cdot \overset{1}{2}}{8 \cdot \underset{1}{2} \cdot b \cdot 1}$$

$$= \dfrac{13}{8b}$$

35. $-\dfrac{15}{32y} \div \dfrac{3}{4} = -\dfrac{15}{32y} \cdot \dfrac{4}{3}$

$$= -\dfrac{15 \cdot 4}{32y \cdot 3}$$

$$= -\dfrac{5 \cdot 3 \cdot 4}{8 \cdot 4 \cdot y \cdot 3}$$

$$= -\dfrac{5 \cdot \overset{1}{3} \cdot \overset{1}{4}}{8 \cdot \underset{1}{4} \cdot y \cdot \underset{1}{3}}$$

$$= -\dfrac{5}{8y}$$

37. $a \div \dfrac{a}{b} = a \cdot \dfrac{b}{a}$

$$= \dfrac{a}{1} \cdot \dfrac{b}{a}$$

$$= \dfrac{a \cdot b}{1 \cdot a}$$

$$= \dfrac{\overset{1}{a} \cdot b}{1 \cdot \underset{1}{a}}$$

$$= \dfrac{b}{1}$$

$$= b$$

39. $\dfrac{x}{y} \div \dfrac{x}{y} = \dfrac{x}{y} \cdot \dfrac{y}{x}$

$$= \dfrac{x \cdot y}{y \cdot x}$$

$$= \dfrac{\overset{1}{x} \cdot \overset{1}{y}}{\underset{1}{y} \cdot \underset{1}{x}}$$

$$= \dfrac{1}{1}$$

$$= 1$$

41. $\dfrac{2s}{3t} \div (-6) = \dfrac{2s}{3t}\left(-\dfrac{1}{6}\right)$

$$= -\dfrac{2s \cdot 1}{3t \cdot 6}$$

$$= -\dfrac{2 \cdot s \cdot 1}{3 \cdot t \cdot 3 \cdot 2}$$

$$= -\dfrac{\overset{1}{2} \cdot s \cdot 1}{3 \cdot t \cdot 3 \cdot \underset{1}{2}}$$

$$= -\dfrac{s}{9t}$$

43.
$$-\frac{9}{8}x \div \frac{3}{4x^2} = -\frac{9x}{8} \cdot \frac{4x^2}{3}$$

$$= -\frac{9x \cdot 4x^2}{8 \cdot 3}$$

$$= -\frac{3 \cdot 3 \cdot x \cdot 4 \cdot x^2}{4 \cdot 2 \cdot 3}$$

$$= -\frac{\overset{1}{\cancel{3}} \cdot 3 \cdot x \cdot \overset{1}{\cancel{4}} \cdot x^2}{4 \cdot 2 \cdot \underset{1}{\cancel{3}}}$$

$$= -\frac{3x^3}{2}$$

47.
$$-\frac{x^2}{y^3} \div \frac{x}{y} = -\frac{x^2}{y^3} \cdot \frac{y}{x}$$

$$= -\frac{x^2 \cdot y}{y^3 \cdot x}$$

$$= -\frac{x \cdot x \cdot y}{y \cdot y \cdot y \cdot x}$$

$$= -\frac{x \cdot \overset{1}{\cancel{x}} \cdot \overset{1}{\cancel{y}}}{y \cdot y \cdot \underset{1}{\cancel{y}} \cdot \underset{1}{\cancel{x}}}$$

$$= -\frac{x}{y^2}$$

45.
$$-8x \div \left(-\frac{4x^3}{9}\right) = -8x\left(-\frac{9}{4x^3}\right)$$

$$= -\frac{8x}{1}\left(-\frac{9}{4x^3}\right)$$

$$= \frac{8x \cdot 9}{1 \cdot 4x^3}$$

$$= \frac{4 \cdot 2 \cdot x \cdot 9}{1 \cdot 4 \cdot x \cdot x \cdot x}$$

$$= \frac{4 \cdot 2 \cdot \overset{1}{\cancel{x}} \cdot 9}{1 \cdot 4 \cdot \underset{1}{\cancel{x}} \cdot x \cdot x}$$

$$= \frac{18}{x^2}$$

49.
$$-\frac{26x}{15} \div \frac{13}{45x} = -\frac{26x}{15} \cdot \frac{45x}{13}$$
$$= -\frac{26x \cdot 45x}{15 \cdot 13}$$
$$= -\frac{13 \cdot 2 \cdot x \cdot 15 \cdot 3 \cdot x}{15 \cdot 13}$$
$$= -\frac{\overset{1}{\cancel{13}} \cdot 2 \cdot x \cdot \overset{1}{\cancel{15}} \cdot 3 \cdot x}{\underset{1}{\cancel{15}} \cdot \underset{1}{\cancel{13}}}$$
$$= -\frac{6x^2}{1}$$
$$= -6x^2$$

APPLICATIONS

51.
$$26 \div \frac{1}{4} = 26 \cdot \frac{4}{1}$$
$$= \frac{26}{1} \cdot \frac{4}{1}$$
$$= \frac{26 \cdot 4}{1 \cdot 1}$$
$$= \frac{104}{1}$$
$$= 104$$

A runner would have to complete 104 laps.

53.
$$\frac{7}{8} \div \frac{1}{64} = \frac{7}{8} \cdot \frac{64}{1}$$
$$= \frac{7 \cdot 64}{8 \cdot 1}$$
$$= \frac{7 \cdot 8 \cdot 8}{8 \cdot 1}$$
$$= \frac{7 \cdot \overset{1}{8} \cdot 8}{\underset{1}{8} \cdot 1}$$
$$= \frac{56}{1}$$
$$= 56$$

Fifty-six $\frac{1}{64}$–inch–wide slices can be cut.

55.
$$(7+8) \div \frac{3}{5} = 15 \div \frac{3}{5}$$
$$= 15 \cdot \frac{5}{3}$$
$$= \frac{15}{1} \cdot \frac{5}{3}$$
$$= \frac{15 \cdot 5}{1 \cdot 3}$$
$$= \frac{5 \cdot 3 \cdot 5}{1 \cdot 3}$$
$$= \frac{5 \cdot \overset{1}{3} \cdot 5}{1 \cdot \underset{1}{3}}$$
$$= \frac{25}{1}$$
$$= 25$$

$$12 \div \frac{2}{5} = 12 \cdot \frac{5}{2}$$
$$= \frac{12}{1} \cdot \frac{5}{2}$$
$$= \frac{12 \cdot 5}{1 \cdot 2}$$
$$= \frac{6 \cdot 2 \cdot 5}{1 \cdot 2}$$
$$= \frac{6 \cdot \overset{1}{2} \cdot 5}{1 \cdot \underset{1}{2}}$$
$$= \frac{30}{1}$$
$$= 30$$

Route 1 takes 25 days while Route 2 takes 30 days so Route 1 will require the fewest days.

57. $6,284 \div \frac{4}{5} = 7,855$

There are 7,855 sections.

REVIEW

63.
$$4x + (-2) = -18$$
$$4x + (-2) + 2 = -18 + 2$$
$$4x = -16$$
$$\frac{4x}{4} = \frac{-16}{4}$$
$$x = -4$$

65. $p = r - c$

67. False

69. $-3t + (-5T) + 4T + 8t = -T + 5t$

STUDY SET Section 4.4

VOCABULARY

1. The <u>least</u> common denominator for a set of fractions is the smallest number each denominator will divide exactly.

3. To express a fraction in <u>higher</u> terms, we multiply the numerator and denominator by the <u>same</u> number.

CONCEPTS

5. a. This rule tells us how to add fractions having like denominators.

   b. To find the sum, it says to add the <u>numerators</u> and then write that result over the <u>common</u> denominator.

7.
$$\frac{3}{8} + \frac{2}{8} = \frac{3+2}{8}$$
$$= \frac{5}{8}$$

9. The denominators are unlike.

11. The numerator and denominator are being multiplied by 4.

13. The LCD is $2 \cdot 2 \cdot 3 \cdot 5 = 60$

15. $5 = \frac{5}{1}$

NOTATION

17.
$$\frac{2}{5} + \frac{1}{3} = \frac{2 \cdot 3}{5 \cdot 3} + \frac{1 \cdot 5}{3 \cdot 5}$$
$$= \frac{6}{15} + \frac{5}{15}$$
$$= \frac{6+5}{15}$$
$$= \frac{11}{15}$$

PRACTICE

19.
$$\frac{3}{4} = \frac{3 \cdot 3}{4 \cdot 3}$$
$$= \frac{9}{12}$$

21.
$$5 = \frac{5}{1}$$
$$= \frac{5 \cdot 2}{1 \cdot 2}$$
$$= \frac{10}{2}$$

23. $\dfrac{1}{t} = \dfrac{1 \cdot 4}{t \cdot 4}$

$\phantom{\dfrac{1}{t}} = \dfrac{4}{4t}$

25. $\dfrac{5}{y} = \dfrac{5 \cdot 6}{y \cdot 6}$

$\phantom{\dfrac{5}{y}} = \dfrac{30}{6y}$

27. $\dfrac{3}{7} + \dfrac{1}{7} = \dfrac{3+1}{7}$

$\phantom{\dfrac{3}{7} + \dfrac{1}{7}} = \dfrac{4}{7}$

29. $\dfrac{37}{103} - \dfrac{17}{103} = \dfrac{37-17}{103}$

$\phantom{\dfrac{37}{103} - \dfrac{17}{103}} = \dfrac{20}{103}$

31. $\dfrac{1}{4} + \dfrac{3}{8} = \dfrac{1 \cdot 2}{4 \cdot 2} + \dfrac{3}{8}$

$\phantom{\dfrac{1}{4} + \dfrac{3}{8}} = \dfrac{2}{8} + \dfrac{3}{8}$

$\phantom{\dfrac{1}{4} + \dfrac{3}{8}} = \dfrac{2+3}{8}$

$\phantom{\dfrac{1}{4} + \dfrac{3}{8}} = \dfrac{5}{8}$

33. $\dfrac{4}{5} - \dfrac{1}{2} = \dfrac{4 \cdot 2}{5 \cdot 2} - \dfrac{1 \cdot 5}{2 \cdot 5}$

$\phantom{\dfrac{4}{5} - \dfrac{1}{2}} = \dfrac{8}{10} - \dfrac{5}{10}$

$\phantom{\dfrac{4}{5} - \dfrac{1}{2}} = \dfrac{8-5}{10}$

$\phantom{\dfrac{4}{5} - \dfrac{1}{2}} = \dfrac{3}{10}$

35. $\dfrac{4}{5} + \dfrac{2}{3} = \dfrac{4 \cdot 3}{5 \cdot 3} + \dfrac{2 \cdot 5}{3 \cdot 5}$

$\phantom{\dfrac{4}{5} + \dfrac{2}{3}} = \dfrac{12}{15} + \dfrac{10}{15}$

$\phantom{\dfrac{4}{5} + \dfrac{2}{3}} = \dfrac{12+10}{15}$

$\phantom{\dfrac{4}{5} + \dfrac{2}{3}} = \dfrac{22}{15}$

37. $-\dfrac{5}{8} - \dfrac{1}{3} = \dfrac{-5}{8} - \dfrac{1}{3}$

$\phantom{-\dfrac{5}{8} - \dfrac{1}{3}} = \dfrac{-5 \cdot 3}{8 \cdot 3} - \dfrac{1 \cdot 8}{3 \cdot 8}$

$\phantom{-\dfrac{5}{8} - \dfrac{1}{3}} = \dfrac{-15}{24} - \dfrac{8}{24}$

$\phantom{-\dfrac{5}{8} - \dfrac{1}{3}} = \dfrac{-15-8}{24}$

$\phantom{-\dfrac{5}{8} - \dfrac{1}{3}} = \dfrac{-23}{24}$

$\phantom{-\dfrac{5}{8} - \dfrac{1}{3}} = -\dfrac{23}{24}$

39. $\dfrac{16}{25} - \left(-\dfrac{3}{10}\right) = \dfrac{16}{25} + \dfrac{3}{10}$

$\phantom{\dfrac{16}{25} - \left(-\dfrac{3}{10}\right)} = \dfrac{16 \cdot 2}{25 \cdot 2} + \dfrac{3 \cdot 5}{10 \cdot 5}$

$\phantom{\dfrac{16}{25} - \left(-\dfrac{3}{10}\right)} = \dfrac{32}{50} + \dfrac{15}{50}$

$\phantom{\dfrac{16}{25} - \left(-\dfrac{3}{10}\right)} = \dfrac{32+15}{50}$

$\phantom{\dfrac{16}{25} - \left(-\dfrac{3}{10}\right)} = \dfrac{47}{50}$

41.

$$-\frac{7}{16} - \frac{1}{4} = \frac{-7}{16} - \frac{1}{4}$$

$$= \frac{-7}{16} - \frac{1 \cdot 4}{4 \cdot 4}$$

$$= \frac{-7}{16} - \frac{4}{16}$$

$$= \frac{-7 - 4}{16}$$

$$= \frac{-11}{16}$$

$$= -\frac{11}{16}$$

43.

$$\frac{4}{7} - \frac{1}{r} = \frac{4 \cdot r}{7 \cdot r} - \frac{1 \cdot 7}{r \cdot 7}$$

$$= \frac{4r}{7r} - \frac{7}{7r}$$

$$= \frac{4r - 7}{7r}$$

45.

$$-\frac{5}{9} + \frac{1}{y} = \frac{-5}{9} + \frac{1}{y}$$

$$= \frac{-5 \cdot y}{9 \cdot y} + \frac{1 \cdot 9}{y \cdot 9}$$

$$= \frac{-5y}{9y} + \frac{9}{9y}$$

$$= \frac{-5y + 9}{9y}$$

47.

$$\frac{7}{8} - \frac{t}{7} = \frac{7 \cdot 7}{8 \cdot 7} - \frac{t \cdot 8}{7 \cdot 8}$$

$$= \frac{49}{56} - \frac{8t}{56}$$

$$= \frac{49 - 8t}{56}$$

49.

$$\frac{4}{5} - \frac{2b}{9} = \frac{4 \cdot 9}{5 \cdot 9} - \frac{2b \cdot 5}{9 \cdot 5}$$

$$= \frac{36}{45} - \frac{10b}{45}$$

$$= \frac{36 - 10b}{45}$$

51.

$$-3 + \frac{2}{5} = \frac{-3}{1} + \frac{2}{5}$$

$$= \frac{-3 \cdot 5}{1 \cdot 5} + \frac{2}{5}$$

$$= \frac{-15}{5} + \frac{2}{5}$$

$$= \frac{-15 + 2}{5}$$

$$= \frac{-13}{5}$$

$$= -\frac{13}{5}$$

53.

$$-\frac{3}{4} - 5 = \frac{-3}{4} - \frac{5}{1}$$

$$= \frac{-3}{4} - \frac{5 \cdot 4}{1 \cdot 4}$$

$$= \frac{-3}{4} - \frac{20}{4}$$

$$= \frac{-3 - 20}{4}$$

$$= \frac{-23}{4}$$

$$= -\frac{23}{4}$$

55. $\dfrac{1}{3}+\dfrac{1}{4}+\dfrac{1}{5}=\dfrac{1\cdot 20}{3\cdot 20}+\dfrac{1\cdot 15}{4\cdot 15}+\dfrac{1\cdot 12}{5\cdot 12}$

$=\dfrac{20}{60}+\dfrac{15}{60}+\dfrac{12}{60}$

$=\dfrac{20+15+12}{60}$

$=\dfrac{47}{60}$

57. $-\dfrac{2}{3}+\dfrac{5}{4}+\dfrac{1}{6}=\dfrac{-2}{3}+\dfrac{5}{4}+\dfrac{1}{6}$

$=\dfrac{-2\cdot 4}{3\cdot 4}+\dfrac{5\cdot 3}{4\cdot 3}+\dfrac{1\cdot 2}{6\cdot 2}$

$=\dfrac{-8}{12}+\dfrac{15}{12}+\dfrac{2}{12}$

$=\dfrac{-8+15+2}{12}$

$=\dfrac{9}{12}$

$=\dfrac{3\cdot \overset{1}{\cancel{3}}}{4\cdot \underset{1}{\cancel{3}}}$

$=\dfrac{3}{4}$

59. $\dfrac{5}{24}+\dfrac{3}{16}=\dfrac{5\cdot 2}{24\cdot 2}+\dfrac{3\cdot 3}{16\cdot 3}$

$=\dfrac{10}{48}+\dfrac{9}{48}$

$=\dfrac{10+9}{48}$

$=\dfrac{19}{48}$

61. $-\dfrac{11}{15}-\dfrac{2}{9}=\dfrac{-11}{15}-\dfrac{2}{9}$

$=\dfrac{-11\cdot 3}{15\cdot 3}-\dfrac{2\cdot 5}{9\cdot 5}$

$=\dfrac{-33}{45}-\dfrac{10}{45}$

$=\dfrac{-33-10}{45}$

$=\dfrac{-43}{45}$

$=-\dfrac{43}{45}$

63. $\dfrac{7}{25}+\dfrac{1}{15}=\dfrac{7\cdot 3}{25\cdot 3}+\dfrac{1\cdot 5}{15\cdot 5}$

$=\dfrac{21}{75}+\dfrac{5}{75}$

$=\dfrac{21+5}{75}$

$=\dfrac{26}{75}$

65. $\dfrac{4}{27}+\dfrac{1}{6}=\dfrac{4\cdot 2}{27\cdot 2}+\dfrac{1\cdot 9}{6\cdot 9}$

$=\dfrac{8}{54}+\dfrac{9}{54}$

$=\dfrac{8+9}{54}$

$=\dfrac{17}{54}$

67. $\dfrac{11}{60} - \dfrac{2}{45} = \dfrac{11 \cdot 3}{60 \cdot 3} - \dfrac{2 \cdot 4}{45 \cdot 4}$

$\qquad\qquad = \dfrac{33}{180} - \dfrac{8}{180}$

$\qquad\qquad = \dfrac{33 - 8}{180}$

$\qquad\qquad = \dfrac{25}{180}$

$\qquad\qquad = \dfrac{5 \cdot \overset{1}{\cancel{5}}}{36 \cdot \underset{1}{\cancel{5}}}$

$\qquad\qquad = \dfrac{5}{36}$

APPLICATIONS

69. a. $\dfrac{5}{32} + \dfrac{1}{16} = \dfrac{5}{32} + \dfrac{1 \cdot 2}{16 \cdot 2}$

$\qquad\qquad\quad = \dfrac{5}{32} + \dfrac{2}{32}$

$\qquad\qquad\quad = \dfrac{5 + 2}{32}$

$\qquad\qquad\quad = \dfrac{7}{32}$

The growth was $\dfrac{7}{32}$ in.

b. $\dfrac{5}{32} - \dfrac{1}{16} = \dfrac{5}{32} - \dfrac{1 \cdot 2}{16 \cdot 2}$

$\qquad\qquad\quad = \dfrac{5}{32} - \dfrac{2}{32}$

$\qquad\qquad\quad = \dfrac{5 - 2}{32}$

$\qquad\qquad\quad = \dfrac{3}{32}$

The difference is $\dfrac{3}{32}$ in.

71. $\dfrac{2}{6} + \dfrac{3}{8} = \dfrac{2 \cdot 4}{6 \cdot 4} + \dfrac{3 \cdot 3}{8 \cdot 3}$

$\qquad\qquad = \dfrac{8}{24} + \dfrac{9}{24}$

$\qquad\qquad = \dfrac{8 + 9}{24}$

$\qquad\qquad = \dfrac{17}{24}$

$\dfrac{17}{24}$ of a pizza was left. No, the family could not have been fed with just one pizza.

73. $\dfrac{3}{4} = \dfrac{3 \cdot 4}{4 \cdot 4}$

$\qquad = \dfrac{12}{16}$

The scale reads $\dfrac{11}{16}$ so the scale is off by $\dfrac{1}{16}$ lb; customers would be undercharged.

75. $\dfrac{3}{4} = \dfrac{3 \cdot 10}{4 \cdot 10}$

$\qquad = \dfrac{30}{40}$

$\dfrac{4}{5} = \dfrac{4 \cdot 8}{5 \cdot 8}$

$\qquad = \dfrac{32}{40}$

$\dfrac{5}{8} = \dfrac{5 \cdot 5}{8 \cdot 5}$

$\qquad = \dfrac{25}{40}$

The lengths from longest to shortest: $\dfrac{4}{5}, \dfrac{3}{4}, \dfrac{5}{8}$

77. $\dfrac{2}{5} + \dfrac{3}{10} = \dfrac{2 \cdot 2}{5 \cdot 2} + \dfrac{3}{10}$

$$= \dfrac{4}{10} + \dfrac{3}{10}$$

$$= \dfrac{4 + 3}{10}$$

$$= \dfrac{7}{10}$$

$\dfrac{7}{10}$ of the students study 2 or more hours daily.

79. $\dfrac{1}{2} - \dfrac{1}{3} = \dfrac{1 \cdot 3}{2 \cdot 3} - \dfrac{1 \cdot 2}{3 \cdot 2}$

$$= \dfrac{3}{6} - \dfrac{2}{6}$$

$$= \dfrac{3 - 2}{6}$$

$$= \dfrac{1}{6}$$

The difference in strength is $\dfrac{1}{6}$ hp.

REVIEW

85. $2(2 + x) - 3(x - 1) = 4 + 2x - 3x + 3$

$$= -x + 7$$

87. $x - 5$

89. $\left(c^4\right)^3 = c^{4 \cdot 3}$

$$= c^{12}$$

91. $P = 2l + 2w$

STUDY SET 4.5

VOCABULARY

1. A <u>mixed number</u> is the sum of a whole number and a proper fraction.

3. To <u>graph</u> a number means to locate its position on a number line and highlight it using a heavy dot.

CONCEPTS

5. a. $-5\dfrac{1}{2}$°

b. $-1\dfrac{7}{8}$

7. a. $-2\dfrac{2}{3}$

  b. $-3\dfrac{1}{3}$

9. $-\dfrac{4}{5},\ -\dfrac{2}{5},\ \dfrac{1}{5}$

11.

13.

NOTATION

15.
$$-5\dfrac{1}{4}\cdot 1\dfrac{1}{7}=-\dfrac{21}{4}\cdot \dfrac{\mathbf{8}}{7}$$
$$=-\dfrac{21\cdot \mathbf{8}}{4\cdot 7}$$
$$=-\dfrac{\overset{1}{\cancel{7}}\cdot 3\cdot \overset{1}{\cancel{4}}\cdot 2}{\underset{1}{\cancel{4}}\cdot \underset{1}{\cancel{7}}}$$
$$=-\dfrac{\mathbf{6}}{1}$$
$$=-6$$

PRACTICE

17.
$$\begin{array}{r} 3 \\ 4\overline{)15} \\ \underline{12} \\ 3 \end{array}$$
$$\dfrac{15}{4}=3\dfrac{3}{4}$$

19.
$$\begin{array}{r} 5 \\ 5\overline{)29} \\ \underline{25} \\ 4 \end{array}$$
$$\dfrac{29}{5}=5\dfrac{4}{5}$$

21.
$$\begin{array}{r} 3 \\ 6\overline{)20} \\ \underline{18} \\ 2 \end{array}$$
$$-\dfrac{20}{6}=-3\dfrac{2}{6}=-3\dfrac{1}{3}$$

23.
$$\begin{array}{r} 10 \\ 12\overline{)127} \\ \underline{12} \\ 07 \end{array}$$
$$\dfrac{127}{12}=10\dfrac{7}{12}$$

25.
$$6\dfrac{1}{2}=\dfrac{6(2)+1}{2}$$
$$=\dfrac{12+1}{2}$$
$$=\dfrac{13}{2}$$

27.
$$20\dfrac{4}{5}=\dfrac{20(5)+4}{5}$$
$$=\dfrac{100+4}{5}$$
$$=\dfrac{104}{5}$$

29.
$$-6\frac{2}{9} = -\frac{6(9)+2}{9}$$
$$= -\frac{54+2}{9}$$
$$= -\frac{56}{9}$$

31.
$$200\frac{2}{3} = \frac{200(3)+2}{3}$$
$$= \frac{600+2}{3}$$
$$= \frac{602}{3}$$

33.

35.

37.
$$1\frac{2}{3} \cdot 2\frac{1}{7} = \frac{5}{3} \cdot \frac{15}{7}$$
$$= \frac{5 \cdot 15}{3 \cdot 7}$$
$$= \frac{5 \cdot 5 \cdot \overset{1}{\cancel{3}}}{\underset{1}{\cancel{3}} \cdot 7}$$
$$= \frac{25}{7}$$
$$= 3\frac{4}{7}$$

39.
$$-7\frac{1}{2}\left(-1\frac{2}{5}\right) = -\frac{15}{2} \cdot -\frac{7}{5}$$
$$= \frac{15 \cdot 7}{2 \cdot 5}$$
$$= \frac{\overset{1}{\cancel{5}} \cdot 3 \cdot 7}{2 \cdot \underset{1}{\cancel{5}}}$$
$$= \frac{21}{2}$$
$$= 10\frac{1}{2}$$

41.
$$3\frac{1}{16} \cdot 4\frac{4}{7} = \frac{49}{16} \cdot \frac{32}{7}$$
$$= \frac{49 \cdot 32}{16 \cdot 7}$$
$$= \frac{\overset{1}{\cancel{7}} \cdot 7 \cdot \overset{1}{\cancel{16}} \cdot 2}{\underset{1}{\cancel{16}} \cdot \underset{1}{\cancel{7}}}$$
$$= \frac{14}{1}$$
$$= 14$$

43.
$$-6 \cdot 2\frac{7}{24} = -\frac{6}{1} \cdot \frac{55}{24}$$
$$= -\frac{6 \cdot 55}{1 \cdot 24}$$
$$= -\frac{\overset{1}{\cancel{6}} \cdot 55}{1 \cdot \cancel{6} \cdot 4}$$
$$= -\frac{55}{4}$$
$$= -13\frac{3}{4}$$

45.
$$2\frac{1}{2}\left(-3\frac{1}{3}\right) = \frac{5}{2} \bullet -\frac{10}{3}$$
$$= -\frac{5 \bullet 10}{2 \bullet 3}$$
$$= -\frac{5 \bullet 5 \bullet \overset{1}{\cancel{2}}}{\underset{1}{\cancel{2}} \bullet 3}$$
$$= -\frac{25}{3}$$
$$= -8\frac{1}{3}$$

47.
$$2\frac{5}{8} \bullet \frac{5}{27} = \frac{21}{8} \bullet \frac{5}{27}$$
$$= \frac{21 \bullet 5}{8 \bullet 27}$$
$$= \frac{7 \bullet \overset{1}{\cancel{3}} \bullet 5}{8 \bullet 9 \bullet \underset{1}{\cancel{3}}}$$
$$= \frac{35}{72}$$

49.
$$1\frac{2}{3} \bullet 6 \bullet -\frac{1}{8} = \frac{5}{3} \bullet \frac{6}{1} \bullet -\frac{1}{8}$$
$$= -\frac{5 \bullet 6 \bullet 1}{3 \bullet 1 \bullet 8}$$
$$= -\frac{5 \bullet \overset{1}{\cancel{3}} \bullet \overset{1}{\cancel{2}} \bullet 1}{\underset{1}{\cancel{3}} \bullet 1 \bullet 4 \bullet \underset{1}{\cancel{2}}}$$
$$= -\frac{5}{4}$$
$$= -1\frac{1}{4}$$

51.
$$\left(1\frac{2}{3}\right)^2 = \left(1\frac{2}{3}\right)\left(1\frac{2}{3}\right)$$
$$= \frac{5}{3} \bullet \frac{5}{3}$$
$$= \frac{5 \bullet 5}{3 \bullet 3}$$
$$= \frac{25}{9}$$

53.
$$\left(-1\frac{1}{3}\right)^3 = \left(-1\frac{1}{3}\right)\left(-1\frac{1}{3}\right)\left(-1\frac{1}{3}\right)$$
$$= \left(-\frac{4}{3}\right)\left(-\frac{4}{3}\right)\left(-\frac{4}{3}\right)$$
$$= -\frac{4 \bullet 4 \bullet 4}{3 \bullet 3 \bullet 3}$$
$$= -\frac{64}{27}$$

55.
$$3\frac{1}{3} \div 1\frac{5}{6} = \frac{10}{3} \div \frac{11}{6}$$
$$= \frac{10}{3} \bullet \frac{6}{11}$$
$$= \frac{10 \bullet 6}{3 \bullet 11}$$
$$= \frac{10 \bullet \overset{1}{\cancel{3}} \bullet 2}{\underset{1}{\cancel{3}} \bullet 11}$$
$$= \frac{20}{11}$$
$$= 1\frac{9}{11}$$

57.
$$-6\frac{3}{5} \div 7\frac{1}{3} = -\frac{33}{5} \div \frac{22}{3}$$
$$= -\frac{33}{5} \cdot \frac{3}{22}$$
$$= -\frac{33 \cdot 3}{5 \cdot 22}$$
$$= -\frac{\overset{1}{11} \cdot 3 \cdot 3}{5 \cdot \underset{1}{11} \cdot 2}$$
$$= -\frac{9}{10}$$

59.
$$-20\frac{1}{4} \div \left(-1\frac{11}{16}\right) = -\frac{81}{4} \div \left(-\frac{27}{16}\right)$$
$$= -\frac{81}{4}\left(-\frac{16}{27}\right)$$
$$= \frac{81 \cdot 16}{4 \cdot 27}$$
$$= \frac{\overset{1}{27} \cdot 3 \cdot \overset{1}{4} \cdot 4}{\underset{1}{4} \cdot \underset{1}{27}}$$
$$= \frac{12}{1}$$
$$= 12$$

61.
$$6\frac{1}{4} \div 20 = \frac{25}{4} \div \frac{20}{1}$$
$$= \frac{25}{4} \cdot \frac{1}{20}$$
$$= \frac{25 \cdot 1}{4 \cdot 20}$$
$$= \frac{5 \cdot \overset{1}{5} \cdot 1}{4 \cdot \underset{1}{5} \cdot 4}$$
$$= \frac{5}{16}$$

63.
$$1\frac{2}{3} \div \left(-2\frac{1}{2}\right) = \frac{5}{3} \div \left(-\frac{5}{2}\right)$$
$$= \frac{5}{3}\left(-\frac{2}{5}\right)$$
$$= -\frac{5 \cdot 2}{3 \cdot 5}$$
$$= -\frac{\overset{1}{5} \cdot 2}{3 \cdot \underset{1}{5}}$$
$$= -\frac{2}{3}$$

65.
$$8 \div 3\frac{1}{5} = \frac{8}{1} \div \frac{16}{5}$$
$$= \frac{8}{1} \cdot \frac{5}{16}$$
$$= \frac{8 \cdot 5}{1 \cdot 16}$$
$$= \frac{\overset{1}{8} \cdot 5}{1 \cdot \underset{1}{8} \cdot 2}$$
$$= \frac{5}{2}$$
$$= 2\frac{1}{2}$$

67.
$$-4\frac{1}{2} \div 2\frac{1}{4} = -\frac{9}{2} \div \frac{9}{4}$$
$$= -\frac{9}{2} \cdot \frac{4}{9}$$
$$= -\frac{9 \cdot 4}{2 \cdot 9}$$
$$= -\frac{\overset{1}{9} \cdot \overset{1}{2} \cdot 2}{\underset{1}{2} \cdot \underset{1}{9}}$$
$$= -\frac{2}{1}$$
$$= -2$$

APPLICATIONS

69.
$$20 \cdot 3\frac{1}{5} = \frac{20}{1} \cdot \frac{16}{5}$$
$$= \frac{20 \cdot 16}{1 \cdot 5}$$
$$= \frac{\overset{1}{\cancel{5}} \cdot 4 \cdot 16}{1 \cdot \underset{1}{\cancel{5}}}$$
$$= \frac{64}{1}$$
$$= 64$$

There are 64 calories in a whole package.

71.
$$4\frac{1}{4} \cdot (0.64) = \frac{17}{4} \cdot \frac{64}{10}$$
$$= \frac{17 \cdot 64}{4 \cdot 100}$$
$$= \frac{17 \cdot 4 \cdot \overset{1}{\cancel{4}} \cdot \overset{1}{\cancel{4}}}{\underset{1}{\cancel{4}} \cdot \underset{1}{\cancel{4}} \cdot 25}$$
$$= \frac{68}{25}$$
$$= 2.72$$

The cost is $2.72.

73.
$$900 \div 1\frac{1}{3} = 900 \div \frac{4}{3}$$
$$= \frac{900}{1} \cdot \frac{3}{4}$$
$$= \frac{900 \cdot 3}{1 \cdot 4}$$
$$= \frac{25 \cdot 9 \cdot \overset{1}{\cancel{4}} \cdot 3}{1 \cdot \underset{1}{\cancel{4}}}$$
$$= \frac{675}{1}$$
$$= 675$$

675 lots were created.

75.
$$11 \cdot \frac{1}{4} = \frac{11}{1} \cdot \frac{1}{4}$$
$$= \frac{11 \cdot 1}{1 \cdot 4}$$
$$= \frac{11}{4}$$
$$= 2\frac{3}{4}$$

$$5 \cdot \frac{1}{4} = \frac{5}{1} \cdot \frac{1}{4}$$
$$= \frac{5 \cdot 1}{1 \cdot 4}$$
$$= \frac{5}{4}$$
$$= 1\frac{1}{4}$$

The length is $2\frac{3}{4}$ in.; the width is $1\frac{1}{4}$ in.

77. $A = \dfrac{1}{2}bh$

$A = \dfrac{1}{2} \cdot 8\dfrac{1}{4} \cdot 10\dfrac{1}{3}$

$= \dfrac{1}{2} \cdot \dfrac{33}{4} \cdot \dfrac{31}{3}$

$= \dfrac{1 \cdot 33 \cdot 31}{2 \cdot 4 \cdot 3}$

$= \dfrac{1 \cdot 11 \cdot \overset{1}{\cancel{3}} \cdot 31}{2 \cdot 4 \cdot \underset{1}{\cancel{3}}}$

$= \dfrac{341}{8}$

$= 42\dfrac{5}{8}$

The area is $42\dfrac{5}{8}$ in.^{2}

79. $(105 \cdot 43) \div 7\dfrac{1}{2} = 4515 \div \dfrac{15}{2}$

$= \dfrac{4515}{1} \cdot \dfrac{2}{15}$

$= 602$

There are 602 steps in the stairway.

REVIEW

85. $3^2 \cdot 2^2 = 9 \cdot 4$

$= 36$

87. $8 + 8 + 8 + 8 = 4(8)$

89. To solve for x, division by 2 must be undone.

91. $3t$ and $4T$ are not like terms because the variables are different.

VOCABULARY

1. By the <u>commutative</u> property of addition, we can add numbers in any order.

CONCEPTS

3. $4\dfrac{3}{5} = 4 + \dfrac{3}{5}$

5. $14 + \dfrac{5}{6} + 53 + \dfrac{1}{6} = 14 + 53 + \dfrac{5}{6} + \dfrac{1}{6}$

7. The LCD for $5 \cdot 2$ and $5 \cdot 3$ is $2 \cdot 3 \cdot 5 = 30$.

9. $16\dfrac{12}{8} = 16 + \dfrac{12}{8}$

$= 16 + 1\dfrac{4}{8}$

$= 16 + 1\dfrac{\overset{1}{\cancel{4}}}{\underset{1}{\cancel{4}} \cdot 2}$

$= 16 + 1\dfrac{1}{2}$

$= 17\dfrac{1}{2}$

NOTATION

11.

$$70\frac{3}{5}+39\frac{2}{7} = \underline{70}+\frac{3}{5}+\underline{39}+\frac{2}{7}$$

$$= \underline{70}+\underline{39}+\frac{3}{5}+\frac{2}{7}$$

$$= 109+\frac{3}{5}+\frac{2}{7}$$

$$= 109+\frac{3\cdot\mathbf{7}}{5\cdot\mathbf{7}}+\frac{2\cdot\mathbf{5}}{7\cdot\mathbf{5}}$$

$$= 109+\frac{21}{\mathbf{35}}+\frac{10}{\mathbf{35}}$$

$$= 109+\frac{\mathbf{31}}{35}$$

$$= 109\frac{31}{35}$$

PRACTICE

13.

$$2\frac{1}{5}+2\frac{1}{5} = \frac{11}{5}+\frac{11}{5}$$

$$= \frac{22}{5}$$

$$= 4\frac{2}{5}$$

15.

$$8\frac{2}{7}-3\frac{1}{7} = \frac{58}{7}-\frac{22}{7}$$

$$= \frac{36}{7}$$

$$= 5\frac{1}{7}$$

17.

$$3\frac{1}{4}+4\frac{1}{4} = \frac{13}{4}+\frac{17}{4}$$

$$= \frac{30}{4}$$

$$= 7\frac{2}{4}$$

$$= 7\frac{1}{2}$$

19.

$$4\frac{1}{6}+1\frac{1}{5} = \frac{25}{6}+\frac{6}{5}$$

$$= \frac{25\cdot5}{6\cdot5}+\frac{6\cdot6}{5\cdot6}$$

$$= \frac{125}{30}+\frac{36}{30}$$

$$= \frac{161}{30}$$

$$= 5\frac{11}{30}$$

21.

$$2\frac{1}{2}-1\frac{1}{4} = \frac{5}{2}-\frac{5}{4}$$

$$= \frac{5\cdot2}{2\cdot2}-\frac{5}{4}$$

$$= \frac{10}{4}-\frac{5}{4}$$

$$= \frac{5}{4}$$

$$= 1\frac{1}{4}$$

23. $2\dfrac{5}{6} - 1\dfrac{3}{8} = \dfrac{17}{6} - \dfrac{11}{8}$

$\qquad = \dfrac{17 \cdot 4}{6 \cdot 4} - \dfrac{11 \cdot 3}{8 \cdot 3}$

$\qquad = \dfrac{68}{24} - \dfrac{33}{24}$

$\qquad = \dfrac{35}{24}$

$\qquad = 1\dfrac{11}{24}$

25. $5\dfrac{1}{2} + 3\dfrac{4}{5} = \dfrac{11}{2} + \dfrac{19}{5}$

$\qquad = \dfrac{11 \cdot 5}{2 \cdot 5} + \dfrac{19 \cdot 2}{5 \cdot 2}$

$\qquad = \dfrac{55}{10} + \dfrac{38}{10}$

$\qquad = \dfrac{93}{10}$

$\qquad = 9\dfrac{3}{10}$

27. $7\dfrac{1}{2} - 4\dfrac{1}{7} = \dfrac{15}{2} - \dfrac{29}{7}$

$\qquad = \dfrac{15 \cdot 7}{2 \cdot 7} - \dfrac{29 \cdot 2}{7 \cdot 2}$

$\qquad = \dfrac{105}{14} - \dfrac{58}{14}$

$\qquad = \dfrac{47}{14}$

$\qquad = 3\dfrac{5}{14}$

29. $56\dfrac{2}{5} + 73\dfrac{1}{3} = 56 + \dfrac{2}{5} + 73 + \dfrac{1}{3}$

$\qquad = 56 + 73 + \dfrac{2}{5} + \dfrac{1}{3}$

$\qquad = 129 + \dfrac{2}{5} + \dfrac{1}{3}$

$\qquad = 129 + \dfrac{2 \cdot 3}{5 \cdot 3} + \dfrac{1 \cdot 5}{3 \cdot 5}$

$\qquad = 129 + \dfrac{6}{15} + \dfrac{5}{15}$

$\qquad = 129 + \dfrac{11}{15}$

$\qquad = 129\dfrac{11}{15}$

31. $380\dfrac{1}{6} + 17\dfrac{1}{4} = 380 + \dfrac{1}{6} + 17 + \dfrac{1}{4}$

$\qquad = 380 + 17 + \dfrac{1}{6} + \dfrac{1}{4}$

$\qquad = 397 + \dfrac{1}{6} + \dfrac{1}{4}$

$\qquad = 397 + \dfrac{1 \cdot 4}{6 \cdot 4} + \dfrac{1 \cdot 6}{4 \cdot 6}$

$\qquad = 397 + \dfrac{4}{24} + \dfrac{6}{24}$

$\qquad = 397 + \dfrac{10}{24}$

$\qquad = 397 + \dfrac{5}{12}$

$\qquad = 397\dfrac{5}{12}$

33.
$$28\frac{5}{9} + 44\frac{2}{3} = 28 + \frac{5}{9} + 44 + \frac{2}{3}$$
$$= 28 + 44 + \frac{5}{9} + \frac{2}{3}$$
$$= 72 + \frac{5}{9} + \frac{2}{3}$$
$$= 72 + \frac{5}{9} + \frac{2\cdot 3}{3\cdot 3}$$
$$= 72 + \frac{5}{9} + \frac{6}{9}$$
$$= 72 + \frac{11}{9}$$
$$= 72 + 1\frac{2}{9}$$
$$= 73\frac{2}{9}$$

35.
$$78\frac{5}{7} = 78\frac{5\cdot 3}{7\cdot 3}$$
$$-55\frac{1}{3} = -55\frac{1\cdot 7}{3\cdot 7}$$

$$78\frac{15}{21}$$
$$-55\frac{7}{21}$$
$$\overline{23\frac{8}{21}}$$

37.
$$40\frac{5}{6} = 40\frac{5\cdot 5}{6\cdot 5}$$
$$-29\frac{4}{5} = -29\frac{4\cdot 6}{5\cdot 6}$$

$$40\frac{25}{30}$$
$$-29\frac{24}{30}$$
$$\overline{11\frac{1}{30}}$$

39.
$$22\frac{13}{16} = 22\frac{13}{16}$$
$$-21\frac{3}{8} = -21\frac{3\cdot 2}{8\cdot 2}$$

$$22\frac{13}{16}$$
$$-21\frac{6}{16}$$
$$\overline{1\frac{7}{16}}$$

41.
$$16\frac{1}{4} = 15\frac{1}{4} + \frac{4}{4}$$
$$-13\frac{3}{4} = -13\frac{3}{4}$$

$$15\frac{5}{4}$$
$$-13\frac{3}{4}$$
$$\overline{2\frac{2}{4} = 2\frac{1}{2}}$$

43.

$$76\frac{1}{6} = 76\frac{1 \cdot 4}{6 \cdot 4}$$
$$\underline{-49\frac{7}{8} = -49\frac{7 \cdot 3}{8 \cdot 3}}$$

$$76\frac{4}{24} = 75\frac{4}{24} + \frac{24}{24}$$
$$\underline{-49\frac{21}{24} = -49\frac{21}{24}}$$

$$75\frac{28}{24}$$
$$\underline{-49\frac{21}{24}}$$
$$26\frac{7}{24}$$

47.

$$34\frac{1}{9} = 34\frac{1 \cdot 2}{9 \cdot 2}$$
$$\underline{-13\frac{5}{6} = -13\frac{5 \cdot 3}{6 \cdot 3}}$$

$$34\frac{2}{18} = 33\frac{2}{18} + \frac{18}{18}$$
$$\underline{-13\frac{15}{18} = -13\frac{15}{18}}$$

$$33\frac{20}{18}$$
$$\underline{-13\frac{15}{18}}$$
$$20\frac{5}{18}$$

45.

$$40\frac{3}{16} = 40\frac{3}{16}$$
$$\underline{-29\frac{3}{4} = -29\frac{3 \cdot 4}{4 \cdot 4}}$$

$$40\frac{3}{16} = 39\frac{3}{16} + \frac{16}{16}$$
$$\underline{-29\frac{12}{16} = -29\frac{12}{16}}$$

$$39\frac{19}{16}$$
$$\underline{-29\frac{12}{16}}$$
$$10\frac{7}{16}$$

49.
$$7 - \frac{2}{3} = \frac{7}{1} - \frac{2}{3}$$
$$= \frac{7 \cdot 3}{1 \cdot 3} - \frac{2}{3}$$
$$= \frac{21}{3} - \frac{2}{3}$$
$$= \frac{19}{3}$$
$$= 6\frac{1}{3}$$

51.
$$9 - 8\frac{3}{4} = \frac{9}{1} - \frac{35}{4}$$
$$= \frac{9 \cdot 4}{1 \cdot 4} - \frac{35}{4}$$
$$= \frac{36}{4} - \frac{35}{4}$$
$$= \frac{1}{4}$$

53. $4\dfrac{1}{7} - \dfrac{4}{5} = \dfrac{29}{7} - \dfrac{4}{5}$

$\qquad = \dfrac{29 \cdot 5}{7 \cdot 5} - \dfrac{4 \cdot 7}{5 \cdot 7}$

$\qquad = \dfrac{145}{35} - \dfrac{28}{35}$

$\qquad = \dfrac{117}{35}$

$\qquad = 3\dfrac{12}{35}$

55. $6\dfrac{5}{8} - 3 = \dfrac{53}{8} - \dfrac{3}{1}$

$\qquad = \dfrac{53}{8} - \dfrac{3 \cdot 8}{1 \cdot 8}$

$\qquad = \dfrac{53}{8} - \dfrac{24}{8}$

$\qquad = \dfrac{29}{8}$

$\qquad = 3\dfrac{5}{8}$

57. $\dfrac{7}{3} + 2 = 2\dfrac{1}{3} + 2$

$\qquad = 2 + \dfrac{1}{3} + 2$

$\qquad = 4 + \dfrac{1}{3}$

$\qquad = 4\dfrac{1}{3}$

59. $2 + 1\dfrac{7}{8} = 2 + 1 + \dfrac{7}{8}$

$\qquad = 3 + \dfrac{7}{8}$

$\qquad = 3\dfrac{7}{8}$

61.

$$
\begin{array}{lll}
12\dfrac{1}{2} = & 12\dfrac{1 \cdot 6}{2 \cdot 6} = & 12\dfrac{6}{12} \\[2mm]
5\dfrac{3}{4} = & 5\dfrac{3 \cdot 3}{4 \cdot 3} = & 5\dfrac{9}{12} \\[2mm]
+35\dfrac{1}{6} = & +35\dfrac{1 \cdot 2}{6 \cdot 2} = & +35\dfrac{2}{12} \\[2mm]
\hline
& & 52\dfrac{17}{12}
\end{array}
$$

$$52\dfrac{17}{12} = 52 + 1\dfrac{5}{12} = 53\dfrac{5}{12}$$

63.

$$
\begin{array}{lll}
58\dfrac{7}{8} = & 58\dfrac{7}{8} = & 58\dfrac{7}{8} \\[2mm]
40 = & 40 = & 40 \\[2mm]
+61\dfrac{1}{4} = & +61\dfrac{1 \cdot 2}{4 \cdot 2} = & +61\dfrac{2}{8} \\[2mm]
\hline
& & 159\dfrac{9}{8}
\end{array}
$$

$$159\dfrac{9}{8} = 159 + 1\dfrac{1}{8} = 160\dfrac{1}{8}$$

65. $-3\dfrac{3}{4} + \left(-1\dfrac{1}{2}\right) = -\dfrac{15}{4} + \left(-\dfrac{3}{2}\right)$

$\qquad = -\dfrac{15}{4} + \left(-\dfrac{3 \cdot 2}{2 \cdot 2}\right)$

$\qquad = -\dfrac{15}{4} + \left(-\dfrac{6}{4}\right)$

$\qquad = -\dfrac{21}{4}$

$\qquad = -5\dfrac{1}{4}$

67.
$$-4\frac{5}{8} - 1\frac{1}{4} = -\frac{37}{8} - \frac{5}{4}$$
$$= -\frac{37}{8} - \frac{5 \cdot 2}{4 \cdot 2}$$
$$= -\frac{37}{8} - \frac{10}{8}$$
$$= -\frac{37}{8} + \left(-\frac{10}{8}\right)$$
$$= -\frac{47}{8}$$
$$= -5\frac{7}{8}$$

APPLICATIONS

69.
$$3\frac{1}{2} - \frac{3}{4} = \frac{7}{2} - \frac{3}{4}$$
$$= \frac{7 \cdot 2}{2 \cdot 2} - \frac{3}{4}$$
$$= \frac{14}{4} - \frac{3}{4}$$
$$= \frac{11}{4}$$
$$= 2\frac{3}{4}$$

You would have to travel $2\frac{3}{4}$ mi. farther.

71. $2\dfrac{3}{4}+2\left(\dfrac{1}{2}\right)+\dfrac{2}{3}+1\dfrac{2}{3}=2+\dfrac{3}{4}+1+\dfrac{2}{3}+1+\dfrac{2}{3}$

$$= 2+1+1+\dfrac{3}{4}+\dfrac{2}{3}+\dfrac{2}{3}$$

$$= 4+\dfrac{3\cdot 3}{4\cdot 3}+\dfrac{2\cdot 4}{3\cdot 4}+\dfrac{2\cdot 4}{3\cdot 4}$$

$$= 4+\dfrac{9}{12}+\dfrac{8}{12}+\dfrac{8}{12}$$

$$= 4+\dfrac{25}{12}$$

$$= 4+2\dfrac{1}{12}$$

$$= 6\dfrac{1}{12}$$

The adjusted recipe will yield $6\dfrac{1}{12}$ cups.

73. $50-1\dfrac{1}{2}=\dfrac{50}{1}-\dfrac{3}{2}$

$$= \dfrac{50\cdot 2}{1\cdot 2}-\dfrac{3}{2}$$

$$= \dfrac{100}{2}-\dfrac{3}{2}$$

$$= \dfrac{97}{2}$$

$$= 48\dfrac{1}{2}$$

The hose is $48\dfrac{1}{2}$ ft after the repair.

75. a.

	Speed (mph)	Time traveling (hr)	Distance traveled (mi)
Passenger ship	$16\dfrac{1}{2}$	1	$16\dfrac{1}{2}$
Cargo ship	$5\dfrac{1}{5}$	1	$5\dfrac{1}{5}$

75. b.

$$16\frac{1}{2} + 5\frac{1}{5} = \frac{33}{2} + \frac{26}{5}$$

$$= \frac{33 \cdot 5}{2 \cdot 5} + \frac{26 \cdot 2}{5 \cdot 2}$$

$$= \frac{165}{10} + \frac{52}{10}$$

$$= \frac{217}{10}$$

$$= 21\frac{7}{10}$$

At 1:00 A.M. they are $21\frac{7}{10}$ mi apart.

77. a.

$$159\frac{9}{10}$$
$$\underline{-139\frac{9}{10}}$$
$$20$$

The difference between the least and most expensive self-serve gas is 20¢.

b.

$$179\frac{9}{10} \qquad 169\frac{9}{10} \qquad 189\frac{9}{10}$$
$$\underline{-149\frac{9}{10}} \qquad \underline{-139\frac{9}{10}} \qquad \underline{-159\frac{9}{10}}$$
$$30 \qquad\qquad 30 \qquad\qquad 30$$

Full service is 30¢ more per gallon for all types.

79.

$$511\frac{5}{12} = 511\frac{5}{12}$$
$$\underline{-159\frac{3}{4}} = \underline{-159\frac{3 \cdot 3}{4 \cdot 3}}$$

$$511\frac{5}{12} = 510\frac{5}{12} + \frac{12}{12}$$
$$\underline{-159\frac{9}{12}} = \underline{-159\frac{9}{12}}$$

$$510\frac{17}{12}$$
$$\underline{-159\frac{9}{12}}$$
$$351\frac{8}{12} = 351\frac{2}{3}$$

Before the addition the slide was $351\frac{2}{3}$ ft long.

REVIEW

85.
$$2x - 1 = 3x - 8$$
$$2x - 1 - 2x = 3x - 8 - 2x$$
$$-1 = x - 8$$
$$-1 + 8 = x - 8 + 8$$
$$7 = x$$

87. $8(x+2) + 3(2-x) = 8x + 16 + 6 - 3x$
$$= 5x + 22$$

89. $-2 - (-8) = -2 + [-(-8)]$
$$= -2 + 8$$
$$= 6$$

91. Area measures the amount of surface a figure encloses.

STUDY SET 4.7

VOCABULARY

1. A <u>complex</u> fraction is a fraction whose numerator or denominator, or both, contain one or more <u>fractions</u> or mixed numbers.

CONCEPTS

3. $\dfrac{\dfrac{2}{3}}{\dfrac{1}{5}} = \dfrac{2}{3} \div \dfrac{1}{5}$

5. The common denominator is 15.

7. The result will be negative.

9. The LCD for denominators of 6, 4, and 5 would be 60.

NOTATION

11. $\dfrac{\dfrac{1}{8}}{\dfrac{3}{4}} = \dfrac{1}{8} \div \dfrac{3}{4}$

$= \dfrac{1}{8} \cdot \dfrac{4}{3}$

$= \dfrac{1 \cdot 4}{8 \cdot 3}$

$= \dfrac{1 \cdot \overset{1}{4}}{2 \cdot \underset{1}{4} \cdot 3}$

$= \dfrac{1}{6}$

PRACTICE

13. $\dfrac{2}{3}\left(-\dfrac{1}{4}\right) + \dfrac{1}{2} = -\dfrac{2 \cdot 1}{3 \cdot 4} + \dfrac{1}{2}$

$= -\dfrac{\overset{1}{2} \cdot 1}{3 \cdot \underset{1}{2} \cdot 2} + \dfrac{1}{2}$

$= -\dfrac{1}{6} + \dfrac{1}{2}$

$= -\dfrac{1}{6} + \dfrac{1 \cdot 3}{2 \cdot 3}$

$= -\dfrac{1}{6} + \dfrac{3}{6}$

$= \dfrac{2}{6}$

$= \dfrac{1}{3}$

15. $\dfrac{4}{5} - \left(-\dfrac{1}{3}\right)^2 = \dfrac{4}{5} - \dfrac{1}{9}$

$= \dfrac{4 \cdot 9}{5 \cdot 9} - \dfrac{1 \cdot 5}{9 \cdot 5}$

$= \dfrac{36}{45} - \dfrac{5}{45}$

$= \dfrac{31}{45}$

17. $-4\left(-\dfrac{1}{5}\right)-\left(\dfrac{1}{4}\right)\left(-\dfrac{1}{2}\right)=-\dfrac{4}{1}\left(-\dfrac{1}{5}\right)-\left(\dfrac{1}{4}\right)\left(-\dfrac{1}{2}\right)$

$$=\dfrac{4}{5}-\left(-\dfrac{1}{8}\right)$$

$$=\dfrac{4\cdot 8}{5\cdot 8}-\left(-\dfrac{1\cdot 5}{8\cdot 5}\right)$$

$$=\dfrac{32}{40}-\left(-\dfrac{5}{40}\right)$$

$$=\dfrac{32}{40}+\dfrac{5}{40}$$

$$=\dfrac{37}{40}$$

19. $1\dfrac{3}{5}\left(\dfrac{1}{2}\right)^{2}\left(\dfrac{3}{4}\right)=1\dfrac{3}{5}\left(\dfrac{1}{4}\right)\left(\dfrac{3}{4}\right)$

$$=\left(\dfrac{8}{5}\right)\left(\dfrac{1}{4}\right)\left(\dfrac{3}{4}\right)$$

$$=\dfrac{4\cdot\overset{1}{2}\cdot 1\cdot 3}{5\cdot\underset{1}{4}\cdot\underset{1}{2}\cdot 2}$$

$$=\dfrac{3}{10}$$

21. $\dfrac{7}{8} - \left(\dfrac{4}{5} + 1\dfrac{3}{4}\right) = \dfrac{7}{8} - \left(\dfrac{4}{5} + \dfrac{7}{4}\right)$

$$= \dfrac{7}{8} - \left(\dfrac{4 \cdot 4}{5 \cdot 4} + \dfrac{7 \cdot 5}{4 \cdot 5}\right)$$

$$= \dfrac{7}{8} - \left(\dfrac{16}{20} + \dfrac{35}{20}\right)$$

$$= \dfrac{7}{8} - \dfrac{51}{20}$$

$$= \dfrac{7 \cdot 5}{8 \cdot 5} - \dfrac{51 \cdot 2}{20 \cdot 2}$$

$$= \dfrac{35}{40} - \dfrac{102}{40}$$

$$= -\dfrac{67}{40}$$

$$= -1\dfrac{27}{40}$$

23. $\left(\dfrac{9}{20} \div 2\dfrac{2}{5}\right) + \left(\dfrac{3}{4}\right)^2 = \left(\dfrac{9}{20} \div \dfrac{12}{5}\right) + \dfrac{9}{16}$

$$= \left(\dfrac{9}{20} \cdot \dfrac{5}{12}\right) + \dfrac{9}{16}$$

$$= \left(\dfrac{3 \cdot \cancel{3} \cdot \cancel{5}}{\cancel{5} \cdot 4 \cdot 4 \cdot \cancel{3}}\right) + \dfrac{9}{16}$$

$$= \dfrac{3}{16} + \dfrac{9}{16}$$

$$= \dfrac{12}{16}$$

$$= \dfrac{3}{4}$$

25. $\left(-\dfrac{3}{4}\cdot\dfrac{9}{16}\right)+\left(\dfrac{1}{2}-\dfrac{1}{8}\right)=-\dfrac{3\cdot 9}{4\cdot 16}+\left(\dfrac{1}{2}-\dfrac{1}{8}\right)$

$$=-\dfrac{27}{64}+\left(\dfrac{1\cdot 4}{2\cdot 4}-\dfrac{1}{8}\right)$$

$$=-\dfrac{27}{64}+\left(\dfrac{4}{8}-\dfrac{1}{8}\right)$$

$$=-\dfrac{27}{64}+\dfrac{3}{8}$$

$$=-\dfrac{27}{64}+\dfrac{3\cdot 8}{8\cdot 8}$$

$$=-\dfrac{27}{64}+\dfrac{24}{64}$$

$$=-\dfrac{3}{64}$$

27. $\left(\dfrac{9}{10}-\dfrac{2}{3}\right)\div\left(-\dfrac{1}{5}\right)=\left(\dfrac{9\cdot 3}{10\cdot 3}-\dfrac{2\cdot 10}{3\cdot 10}\right)\div\left(-\dfrac{1}{5}\right)$

$$=\left(\dfrac{27}{30}-\dfrac{20}{30}\right)\div\left(-\dfrac{1}{5}\right)$$

$$=\dfrac{7}{30}\div\left(-\dfrac{1}{5}\right)$$

$$=\dfrac{7}{30}\left(-\dfrac{5}{1}\right)$$

$$=-\dfrac{7\cdot \overset{1}{\cancel{5}}}{6\cdot \underset{1}{\cancel{5}}\cdot 1}$$

$$=-\dfrac{7}{6}$$

$$=-1\dfrac{1}{6}$$

29.
$$\left(2-\frac{1}{2}\right)^2+\left(2+\frac{1}{2}\right)^2=\left(\frac{4}{2}-\frac{1}{2}\right)^2+\left(\frac{4}{2}+\frac{1}{2}\right)^2$$
$$=\left(\frac{3}{2}\right)^2+\left(\frac{5}{2}\right)^2$$
$$=\frac{9}{4}+\frac{25}{4}$$
$$=\frac{34}{4}$$
$$=8\frac{2}{4}$$
$$=8\frac{1}{2}$$

31.
$$\left(\frac{1}{2}\cdot-7\right)^2=\left(\frac{1}{2}\cdot-\frac{7}{1}\right)^2$$
$$=\left(-\frac{7}{2}\right)^2$$
$$=\frac{49}{4}$$

33.
$$\left(\frac{1}{2}\cdot\frac{11}{2}\right)^2=\left(\frac{11}{4}\right)^2$$
$$=\frac{121}{16}$$

35.
$$\frac{1}{3}b^2+c=\frac{1}{3}\left(-\frac{1}{5}\right)^2+\left(-\frac{2}{3}\right)$$
$$=\frac{1}{3}\left(\frac{1}{25}\right)+\left(-\frac{2}{3}\right)$$
$$=\frac{1}{75}+\left(-\frac{2}{3}\right)$$
$$=\frac{1}{75}+\left(-\frac{2\cdot25}{3\cdot25}\right)$$
$$=\frac{1}{75}+\left(-\frac{50}{75}\right)$$
$$=-\frac{49}{75}$$

37.

$$-1 - ar = -1 - \left(1\frac{3}{4}\right)\left(-1\frac{2}{3}\right)$$

$$= -1 - \left(\frac{7}{4}\right)\left(-\frac{5}{3}\right)$$

$$= -1 - \left(-\frac{35}{12}\right)$$

$$= -1 + \frac{35}{12}$$

$$= -\frac{12}{12} + \frac{35}{12}$$

$$= \frac{23}{12}$$

$$= 1\frac{11}{12}$$

39.

$$P = 2l + 2w$$

$$P = 2\left(2\frac{7}{8}\right) + 2\left(1\frac{1}{4}\right)$$

$$= \left(\frac{2}{1} \cdot \frac{23}{8}\right) + \left(\frac{2}{1} \cdot \frac{5}{4}\right)$$

$$= \frac{\overset{1}{2} \cdot 23}{1 \cdot 4 \cdot \underset{1}{2}} + \frac{\overset{1}{2} \cdot 5}{1 \cdot \underset{1}{2} \cdot 2}$$

$$= \frac{23}{4} + \frac{5}{2}$$

$$= \frac{23}{4} + \frac{5 \cdot 2}{2 \cdot 2}$$

$$= \frac{23}{4} + \frac{10}{4}$$

$$= \frac{33}{4}$$

$$= 8\frac{1}{4}$$

The perimeter is $8\frac{1}{4}$ in.

41.

$$\frac{\dfrac{2}{3}}{\dfrac{4}{5}} = \frac{2}{3} \div \frac{4}{5}$$

$$= \frac{2}{3} \cdot \frac{5}{4}$$

$$= \frac{\overset{1}{2} \cdot 5}{3 \cdot \underset{1}{2} \cdot 2}$$

$$= \frac{5}{6}$$

43.

$$\frac{-\dfrac{14}{15}}{\dfrac{7}{10}} = -\frac{14}{15} \div \frac{7}{10}$$

$$= -\frac{14}{15} \cdot \frac{10}{7}$$

$$= -\frac{14 \cdot 10}{15 \cdot 7}$$

$$= -\frac{\overset{1}{7} \cdot 2 \cdot \overset{1}{5} \cdot 2}{\underset{1}{5} \cdot 3 \cdot \underset{1}{7}}$$

$$= -\frac{4}{3}$$

$$= -1\frac{1}{3}$$

45.

$$\frac{\dfrac{5}{10}}{21} = 5 \div \frac{10}{21}$$

$$= \frac{5}{1} \cdot \frac{21}{10}$$

$$= \frac{5 \cdot 21}{1 \cdot 10}$$

$$= \frac{\overset{1}{\cancel{5}} \cdot 21}{1 \cdot \cancel{5} \cdot 2}$$

$$= \frac{21}{2}$$

$$= 10\frac{1}{2}$$

47.

$$\frac{-\dfrac{5}{6}}{-1\dfrac{7}{8}} = -\frac{5}{6} \div \left(-1\frac{7}{8}\right)$$

$$= -\frac{5}{6} \div \left(-\frac{15}{8}\right)$$

$$= -\frac{5}{6}\left(-\frac{8}{15}\right)$$

$$= \frac{5 \cdot 8}{6 \cdot 15}$$

$$= \frac{\overset{1}{\cancel{5}} \cdot 4 \cdot \overset{1}{\cancel{2}}}{3 \cdot \underset{1}{\cancel{2}} \cdot \underset{1}{\cancel{5}} \cdot 3}$$

$$= \frac{4}{9}$$

49.

$$\frac{\dfrac{1}{2}+\dfrac{1}{4}}{\dfrac{1}{2}-\dfrac{1}{4}} = \frac{4\left(\dfrac{1}{2}+\dfrac{1}{4}\right)}{4\left(\dfrac{1}{2}-\dfrac{1}{4}\right)}$$

$$= \frac{4\left(\dfrac{1}{2}\right)+4\left(\dfrac{1}{4}\right)}{4\left(\dfrac{1}{2}\right)-4\left(\dfrac{1}{4}\right)}$$

$$= \frac{2+1}{2-1}$$

$$= \frac{3}{1}$$

$$= 3$$

51.

$$\frac{\dfrac{3}{8}+\dfrac{1}{4}}{\dfrac{3}{8}-\dfrac{1}{4}} = \frac{8\left(\dfrac{3}{8}+\dfrac{1}{4}\right)}{8\left(\dfrac{3}{8}-\dfrac{1}{4}\right)}$$

$$= \frac{8\left(\dfrac{3}{8}\right)+8\left(\dfrac{1}{4}\right)}{8\left(\dfrac{3}{8}\right)-8\left(\dfrac{1}{4}\right)}$$

$$= \frac{3+2}{3-2}$$

$$= \frac{5}{1}$$

$$= 5$$

53.
$$\frac{\dfrac{1}{5}+3}{-\dfrac{4}{25}} = \frac{25\left(\dfrac{1}{5}+3\right)}{25\left(-\dfrac{4}{25}\right)}$$

$$= \frac{25\left(\dfrac{1}{5}\right)+25(3)}{25\left(-\dfrac{4}{25}\right)}$$

$$= \frac{5+75}{-4}$$

$$= \frac{80}{-4}$$

$$= -20$$

55.
$$\frac{5\dfrac{1}{2}}{-\dfrac{1}{4}+\dfrac{3}{4}} = \frac{4\left(\dfrac{11}{2}\right)}{4\left(-\dfrac{1}{4}+\dfrac{3}{4}\right)}$$

$$= \frac{4\left(\dfrac{11}{2}\right)}{4\left(-\dfrac{1}{4}\right)+4\left(\dfrac{3}{4}\right)}$$

$$= \frac{22}{-1+3}$$

$$= \frac{22}{2}$$

$$= 11$$

57.
$$\frac{\dfrac{1}{5}-\left(-\dfrac{1}{4}\right)}{\dfrac{1}{4}+\dfrac{4}{5}} = \frac{\dfrac{1}{5}+\dfrac{1}{4}}{\dfrac{1}{4}+\dfrac{4}{5}}$$

$$= \frac{20\left(\dfrac{1}{5}+\dfrac{1}{4}\right)}{20\left(\dfrac{1}{4}+\dfrac{4}{5}\right)}$$

$$= \frac{20\left(\dfrac{1}{5}\right)+20\left(\dfrac{1}{4}\right)}{20\left(\dfrac{1}{4}\right)+20\left(\dfrac{4}{5}\right)}$$

$$= \frac{4+5}{5+16}$$

$$= \frac{9}{21}$$

$$= \frac{3}{7}$$

59.
$$\frac{\dfrac{1}{3}+\left(-\dfrac{5}{6}\right)}{1\dfrac{1}{3}} = \frac{\dfrac{1}{3}+\left(-\dfrac{5}{6}\right)}{\dfrac{4}{3}}$$

$$= \frac{6\left[\dfrac{1}{3}+\left(-\dfrac{5}{6}\right)\right]}{6\left(\dfrac{4}{3}\right)}$$

$$= \frac{6\left(\dfrac{1}{3}\right)+6\left(-\dfrac{5}{6}\right)}{6\left(\dfrac{4}{3}\right)}$$

$$= \frac{2+(-5)}{8}$$

$$= \frac{-3}{8}$$

61.

$$\frac{x+y}{2} = \frac{-\dfrac{3}{4}+\dfrac{7}{8}}{2}$$

$$= \frac{8\left(-\dfrac{3}{4}+\dfrac{7}{8}\right)}{8(2)}$$

$$= \frac{8\left(-\dfrac{3}{4}\right)+8\left(\dfrac{7}{8}\right)}{8(2)}$$

$$= \frac{-6+7}{16}$$

$$= \frac{1}{16}$$

63.

$$\frac{2x}{y-x} = \frac{2\left(-\dfrac{3}{4}\right)}{\dfrac{7}{8}-\left(-\dfrac{3}{4}\right)}$$

$$= \frac{-\dfrac{3}{2}}{\dfrac{7}{8}+\dfrac{3}{4}}$$

$$= \frac{8\left(-\dfrac{3}{2}\right)}{8\left(\dfrac{7}{8}+\dfrac{3}{4}\right)}$$

$$= \frac{8\left(-\dfrac{3}{2}\right)}{8\left(\dfrac{7}{8}\right)+8\left(\dfrac{3}{4}\right)}$$

$$= \frac{-12}{7+6}$$

$$= -\frac{12}{13}$$

APPLICATIONS

65.

$$\frac{1\dfrac{3}{4}+2\dfrac{1}{2}}{\dfrac{1}{2}} = \frac{\dfrac{7}{4}+\dfrac{5}{2}}{\dfrac{1}{2}}$$

$$= \frac{4\left(\dfrac{7}{4}+\dfrac{5}{2}\right)}{4\left(\dfrac{1}{2}\right)}$$

$$= \frac{4\left(\dfrac{7}{4}\right)+4\left(\dfrac{5}{2}\right)}{4\left(\dfrac{1}{2}\right)}$$

$$= \frac{7+10}{2}$$

$$= \frac{17}{2}$$

$$= 8\dfrac{1}{2}$$

He can make $8\dfrac{1}{2}$ sandwiches.

67.

	Rate (mph)	Time (hr)	Distance (mi)
Jogger	$2\frac{1}{2}$	$1\frac{1}{2}$	$2\frac{1}{2}\left(1\frac{1}{2}\right)$
Cyclist	$7\frac{1}{5}$	$1\frac{1}{2}$	$7\frac{1}{5}\left(1\frac{1}{2}\right)$

$$
2\frac{1}{2}\left(1\frac{1}{2}\right) + 7\frac{1}{5}\left(1\frac{1}{2}\right) = \frac{5}{2}\left(\frac{3}{2}\right) + \frac{36}{5}\left(\frac{3}{2}\right)
$$

$$
= \frac{5\cdot 3}{2\cdot 2} + \frac{36\cdot 3}{5\cdot 2}
$$

$$
= \frac{15}{4} + \frac{18\cdot \overset{1}{2}\cdot 3}{5\cdot \underset{1}{2}}
$$

$$
= \frac{15}{4} + \frac{54}{5}
$$

$$
= \frac{15\cdot 5}{4\cdot 5} + \frac{54\cdot 4}{5\cdot 4}
$$

$$
= \frac{75}{20} + \frac{216}{20}
$$

$$
= \frac{291}{20}
$$

$$
= 14\frac{11}{20}
$$

In $1\frac{1}{2}$ hours they are $14\frac{11}{20}$ mi apart.

69.
$$\frac{1}{16} + \frac{5}{8} + 3\left(\frac{1}{16}\right) = \frac{1}{16} + \frac{5}{8} + \frac{3}{16}$$
$$= \frac{1}{16} + \frac{5 \cdot 2}{8 \cdot 2} + \frac{3}{16}$$
$$= \frac{1}{16} + \frac{10}{16} + \frac{3}{16}$$
$$= \frac{14}{16}$$
$$= \frac{7}{8}$$

Yes, the total weight is $\frac{7}{8}$ oz

which is less than 1 oz.

71.
$$7\left(\frac{1}{4}\right) + 7\left(\frac{1}{2}\right) + 7\left(\frac{3}{4}\right) = \frac{7}{4} + \frac{7}{2} + \frac{21}{4}$$
$$= \frac{7}{4} + \frac{7 \cdot 2}{2 \cdot 2} + \frac{21}{4}$$
$$= \frac{7}{4} + \frac{14}{4} + \frac{21}{4}$$
$$= \frac{42}{4}$$
$$= 10\frac{2}{4}$$
$$= 10\frac{1}{2}$$

The total distance she walked over the three-week period was $10\frac{1}{2}$ miles.

73.

$$\frac{1}{\dfrac{1}{10}+\dfrac{1}{15}}=\frac{30(1)}{30\left(\dfrac{1}{10}+\dfrac{1}{15}\right)}$$

$$=\frac{30(1)}{30\left(\dfrac{1}{10}\right)+30\left(\dfrac{1}{15}\right)}$$

$$=\frac{30}{3+2}$$

$$=\frac{30}{5}$$

$$=6$$

It takes 6 seconds to refill the pool.

REVIEW

79. $\quad -4d-(-7d)=-4d+7d$
$$=3d$$

81. $\quad 2x(-x)$

83. $\quad 2+3\big[-3-(-4-1)\big]=2+3\big[-3-(-5)\big]$
$$=2+3(-3+5)$$
$$=2+3(2)$$
$$=2+6$$
$$=8$$

85. $\quad 3\cdot3\cdot3\cdot x\cdot x\cdot x\cdot x\cdot x=27x^5$

STUDY SET Section 4.8

VOCABULARY

1. To find the <u>reciprocal</u> of a fraction, invert the numerator and the denominator.

3. The <u>least common denominator</u> of a set of fractions is the smallest number each denominator will divide evenly.

CONCEPTS

5. Both sides were not multiplied by 30.

7. When a number is multiplied by its reciprocal, the result is 1.

9. a. $\dfrac{9}{7}$

b. -2

11. Multiply both sides by $\dfrac{3}{2}$, or multiply both sides by 3 and then divide both sides by 2.

NOTATION

13.
$$\dfrac{\dfrac{8}{7}\left(\dfrac{7}{8}x\right) = \dfrac{8}{7}(21)}{x = 24}$$

15.
$$\dfrac{6\left(h + \dfrac{1}{2}\right) = 6\left(\dfrac{2}{3}\right)}{}$$
$$6h + 6\left(\dfrac{1}{2}\right) = 6\left(\dfrac{2}{3}\right)$$
$$6h + \underline{\ 3\ } = 4$$
$$6h + 3 - \underline{\ 3\ } = 4 - \underline{\ 3\ }$$
$$6h = 1$$
$$\dfrac{6h}{6} = \dfrac{1}{6}$$
$$h = \dfrac{1}{6}$$

PRACTICE

17.
$$\dfrac{4}{7}x = 16$$
$$\dfrac{7}{4}\left(\dfrac{4}{7}x\right) = \dfrac{7}{4}(16)$$
$$x = 28$$

19.
$$\dfrac{7}{8}t = -28$$
$$8\left(\dfrac{7}{8}t\right) = 8(-28)$$
$$7t = -224$$
$$\dfrac{7t}{7} = \dfrac{-224}{7}$$
$$t = -32$$

21.
$$-\dfrac{3}{5}h = 4$$
$$-\dfrac{5}{3}\left(-\dfrac{3}{5}h\right) = -\dfrac{5}{3}(4)$$
$$h = -\dfrac{20}{3}$$

23.
$$\dfrac{2}{3}x = \dfrac{4}{5}$$
$$\dfrac{3}{2}\left(\dfrac{2}{3}x\right) = \dfrac{3}{2}\left(\dfrac{4}{5}\right)$$
$$x = \dfrac{3 \cdot \overset{1}{2} \cdot 2}{\underset{1}{2} \cdot 5}$$
$$x = \dfrac{6}{5}$$

25.
$$\dfrac{2}{5}y = 0$$
$$\dfrac{5}{2}\left(\dfrac{2}{5}y\right) = \dfrac{5}{2}(0)$$
$$y = 0$$

27.
$$-\dfrac{5c}{6} = -25$$
$$-\dfrac{6}{5}\left(-\dfrac{5c}{6}\right) = -\dfrac{6}{5}(-25)$$
$$c = 30$$

29.
$$\dfrac{-5f}{7} = -2$$
$$-\dfrac{7}{5}\left(\dfrac{-5f}{7}\right) = -\dfrac{7}{5}(-2)$$
$$f = \dfrac{14}{5}$$

31.

$$\frac{5}{8}y = \frac{1}{10}$$

$$\frac{8}{5}\left(\frac{5}{8}y\right) = \frac{8}{5}\left(\frac{1}{10}\right)$$

$$y = \frac{4 \cdot 2 \cdot \overset{1}{\cancel{1}}}{5 \cdot 5 \cdot \underset{1}{\cancel{2}}}$$

$$y = \frac{4}{25}$$

33.

$$x - \frac{1}{9} = \frac{7}{9}$$

$$x - \frac{1}{9} + \frac{1}{9} = \frac{7}{9} + \frac{1}{9}$$

$$x = \frac{8}{9}$$

35.

$$x + \frac{1}{9} = \frac{4}{9}$$

$$x + \frac{1}{9} - \frac{1}{9} = \frac{4}{9} - \frac{1}{9}$$

$$x = \frac{3}{9}$$

$$x = \frac{1}{3}$$

37.

$$x - \frac{1}{6} = \frac{2}{9}$$

$$x - \frac{1}{6} + \frac{1}{6} = \frac{2}{9} + \frac{1}{6}$$

$$x = \frac{2}{9} + \frac{1}{6}$$

$$x = \frac{2 \cdot 2}{9 \cdot 2} + \frac{1 \cdot 3}{6 \cdot 3}$$

$$x = \frac{4}{18} + \frac{3}{18}$$

$$x = \frac{7}{18}$$

39.

$$y + \frac{7}{8} = \frac{1}{4}$$

$$y + \frac{7}{8} - \frac{7}{8} = \frac{1}{4} - \frac{7}{8}$$

$$y = \frac{1}{4} - \frac{7}{8}$$

$$y = \frac{1 \cdot 2}{4 \cdot 2} - \frac{7}{8}$$

$$y = \frac{2}{8} - \frac{7}{8}$$

$$y = -\frac{5}{8}$$

41.

$$\frac{5}{4} + t = \frac{1}{4}$$

$$\frac{5}{4} + t - \frac{5}{4} = \frac{1}{4} - \frac{5}{4}$$

$$t = -\frac{4}{4}$$

$$t = -1$$

43.

$$x + \frac{3}{4} = -\frac{1}{2}$$

$$x + \frac{3}{4} - \frac{3}{4} = -\frac{1}{2} - \frac{3}{4}$$

$$x = -\frac{1}{2} - \frac{3}{4}$$

$$x = -\frac{1 \cdot 2}{2 \cdot 2} - \frac{3}{4}$$

$$x = -\frac{2}{4} + \left(-\frac{3}{4}\right)$$

$$x = -\frac{5}{4}$$

45.
$$\frac{-x}{4} + 1 = 10$$
$$\frac{-x}{4} + 1 - 1 = 10 - 1$$
$$\frac{-x}{4} = 9$$
$$-4\left(\frac{-x}{4}\right) = -4(9)$$
$$x = -36$$

47.
$$2x - \frac{1}{2} = \frac{1}{3}$$
$$6\left(2x - \frac{1}{2}\right) = 6\left(\frac{1}{3}\right)$$
$$12x - 6\left(\frac{1}{2}\right) = 6\left(\frac{1}{3}\right)$$
$$12x - 3 = 2$$
$$12x - 3 + 3 = 2 + 3$$
$$12x = 5$$
$$\frac{12x}{12} = \frac{5}{12}$$
$$x = \frac{5}{12}$$

49.
$$\frac{1}{2}x - \frac{1}{9} = \frac{1}{3}$$
$$18\left(\frac{1}{2}x - \frac{1}{9}\right) = 18\left(\frac{1}{3}\right)$$
$$18\left(\frac{1}{2}x\right) - 18\left(\frac{1}{9}\right) = 18\left(\frac{1}{3}\right)$$
$$9x - 2 = 6$$
$$9x - 2 + 2 = 6 + 2$$
$$9x = 8$$
$$\frac{9x}{9} = \frac{8}{9}$$
$$x = \frac{8}{9}$$

51.
$$5 + \frac{x}{3} = \frac{1}{2}$$
$$6\left(5 + \frac{x}{3}\right) = 6\left(\frac{1}{2}\right)$$
$$30 + 6\left(\frac{x}{3}\right) = 6\left(\frac{1}{2}\right)$$
$$30 + 2x = 3$$
$$30 + 2x - 30 = 3 - 30$$
$$2x = -27$$
$$\frac{2x}{2} = \frac{-27}{2}$$
$$x = -\frac{27}{2}$$

53.
$$\frac{2}{5}x + 1 = \frac{1}{3} + x$$
$$15\left(\frac{2}{5}x + 1\right) = 15\left(\frac{1}{3} + x\right)$$
$$15\left(\frac{2}{5}x\right) + 15 = 15\left(\frac{1}{3}\right) + 15x$$
$$6x + 15 = 5 + 15x$$
$$6x + 15 - 6x = 5 + 15x - 6x$$
$$15 = 5 + 9x$$
$$15 - 5 = 5 + 9x - 5$$
$$10 = 9x$$
$$\frac{10}{9} = \frac{9x}{9}$$
$$\frac{10}{9} = x$$

55.
$$\frac{x}{3} + \frac{x}{4} = -2$$
$$12\left(\frac{x}{3} + \frac{x}{4}\right) = 12(-2)$$
$$12\left(\frac{x}{3}\right) + 12\left(\frac{x}{4}\right) = -24$$
$$4x + 3x = -24$$
$$7x = -24$$
$$\frac{7x}{7} = \frac{-24}{7}$$
$$x = -\frac{24}{7}$$

57.
$$4 + \frac{s}{3} = 8$$
$$3\left(4 + \frac{s}{3}\right) = 3(8)$$
$$12 + 3\left(\frac{s}{3}\right) = 24$$
$$12 + s = 24$$
$$12 + s - 12 = 24 - 12$$
$$s = 12$$

59.
$$\frac{5h}{6} - 8 = 12$$
$$\frac{5h}{6} - 8 + 8 = 12 + 8$$
$$\frac{5h}{6} = 20$$
$$\frac{6}{5}\left(\frac{5h}{6}\right) = \frac{6}{5}(20)$$
$$h = 24$$

61.
$$-4 + 9 + \frac{5}{12} = 0$$
$$5 + \frac{5t}{12} = 0$$
$$5 + \frac{5t}{12} - 5 = 0 - 5$$
$$\frac{5t}{12} = -5$$
$$\frac{12}{5}\left(\frac{5t}{12}\right) = \frac{12}{5}(-5)$$
$$t = -12$$

63.
$$-3 - 2 + \frac{4x}{15} = 0$$
$$-5 + \frac{4x}{15} = 0$$
$$-5 + \frac{4x}{15} + 5 = 0 + 5$$
$$\frac{4x}{15} = 5$$
$$\frac{15}{4}\left(\frac{4x}{15}\right) = \frac{15}{4}(5)$$
$$x = \frac{75}{4}$$

APPLICATIONS

65. **A:** Find the number of people needing transmission work; 1 unknown

F: Let x = the number of people needing transmission work

Key phrase: one-third of

Translation: multiply

So $\frac{1}{3}x$ = number of new transmissions installed

The number of new transmissions installed	is	32.

$$\frac{1}{3}x \qquad = \qquad \underline{32}$$

S: $\frac{1}{3}x = 32$

$x = \underline{96}$

S: 96 people brought their cars in for transmission work

C: If we find $\frac{1}{3}$ of $\underline{96}$, we get $\underline{32}$. The answer checks.

67. Let x = the number of baby teeth the child will have.

$$\frac{4}{5}x = 16$$

$$\frac{5}{4}\left(\frac{4}{5}x\right) = \frac{5}{4}(16)$$

$$x = 20$$

The baby will have 20 baby teeth.

69. Let x = the number of homes.

$$\frac{3}{4}x = x - 9$$

$$\frac{3}{4}x - x = x - 9 - x$$

$$\frac{3}{4}x - \frac{4}{4}x = -9$$

$$-\frac{1}{4}x = -9$$

$$-4\left(-\frac{1}{4}x\right) = -4(-9)$$

$$x = 36$$

There are 36 homes in the subdivision.

71. Let x = the total number of pages.

$$\frac{2}{3}x = x - 150$$

$$\frac{2}{3}x - x = x - 150 - x$$

$$\frac{2}{3}x - \frac{3}{3}x = -150$$

$$-\frac{1}{3}x = -150$$

$$-3\left(-\frac{1}{3}x\right) = -3(-150)$$

$$x = 450$$

There are 450 pages in the telephone book.

73. Let w = the width.

$$A = lw$$

$$30 = \frac{15}{4}x$$

$$\frac{4}{15}(30) = \frac{4}{15}\left(\frac{15}{4}x\right)$$

$$8 = x$$

The lights must be 8 in. wide.

75. Let x = the length of class.

$$x - \frac{1}{4}x - \frac{2}{3}x = 30$$

$$12\left(x - \frac{1}{4}x - \frac{2}{3}x\right) = 12(30)$$

$$12x - 12\left(\frac{1}{4}x\right) - 12\left(\frac{2}{3}x\right) = 360$$

$$12x - 3x - 8x = 360$$

$$x = 360$$

The class is 360 minutes long.

REVIEW

81. $a(b+c) = \underline{\ ab + ac\ }$

83.
$$C = \frac{5(F-32)}{9}$$

$$C = \frac{5(41-32)}{9}$$

$$= \frac{5(9)}{9}$$

$$= \frac{45}{9}$$

$$= 5$$

So, $41°F = 5°C$

85.
$$5x - 3 = 2x + 12$$

$$5x - 3 - 2x = 2x + 12 - 2x$$

$$3x - 3 = 12$$

$$3x - 3 + 3 = 12 + 3$$

$$3x = 15$$

$$\frac{3x}{3} = \frac{15}{3}$$

$$x = 5$$

87. 13,000,000

KEY CONCEPT The Fundamental Property of Fractions

1. *Step 1*: The numerator and denominator share a common factor of $\underline{\ 5\ }$.

Step 2: Apply the fundamental property of fractions. Divide the numerator and denominator by the common factor.

$$\frac{15}{25} = \frac{15 \div 5}{25 \div 5}$$

Step 3: Do the divisions to simplify the fraction.

$$= \frac{3}{5}$$

3. *Step 1:* We must multiply 3 by $\underline{6x}$ to obtain $18x$.

Step 2: Apply the fundamental property of fractions. Multiply the numerator and denominator by $6x$.

$$\frac{2}{3} = \frac{2 \cdot 6x}{3 \cdot 6x}$$

Step 3: Do the multiplication in the numerator and the denominator.

$$= \frac{12x}{18x}$$

CHAPTER 4 REVIEW

1. She spends $\dfrac{7}{24}$ of a day sleeping.

3. $\dfrac{2}{-3} = -\dfrac{2}{3} = \dfrac{-2}{3}$

5. The numerator and denominator of the fraction are being divided by 2.

7. a.
$$\frac{15}{45} = \frac{15 \div 15}{45 \div 15}$$
$$= \frac{1}{3}$$

b.
$$\frac{20}{48} = \frac{20 \div 4}{48 \div 4}$$
$$= \frac{5}{12}$$

c.
$$\frac{63x^2}{84x} = \frac{\overset{1}{7} \cdot \overset{1}{3} \cdot 3 \cdot \overset{1}{x} \cdot x}{\underset{1}{7} \cdot 4 \cdot \underset{1}{3} \cdot \underset{1}{x}}$$
$$= \frac{3x}{4}$$

d.
$$\frac{66m^3n}{108m^4n} = \frac{11 \cdot \overset{1}{6} \cdot \overset{1}{m} \cdot \overset{1}{m} \cdot \overset{1}{m} \cdot \overset{1}{n}}{18 \cdot \underset{1}{6} \cdot \underset{1}{m} \cdot \underset{1}{m} \cdot \underset{1}{m} \cdot m \cdot \underset{1}{n}}$$
$$= \frac{11}{18m}$$

9. The numerator and denominator of the original fraction are being multiplied by 2 to obtain an equivalent fraction in higher terms.

11. a. $\dfrac{1}{2} \cdot \dfrac{1}{3} = \dfrac{1 \cdot 1}{2 \cdot 3}$

$= \dfrac{1}{6}$

b. $\dfrac{2}{5}\left(-\dfrac{7}{9}\right) = -\dfrac{2 \cdot 7}{5 \cdot 9}$

$= -\dfrac{14}{45}$

c. $\dfrac{9}{16} \cdot \dfrac{20}{27} = \dfrac{9 \cdot 20}{16 \cdot 27}$

$= \dfrac{\overset{1}{\cancel{9}} \cdot 5 \cdot \overset{1}{\cancel{4}}}{4 \cdot \underset{1}{\cancel{4}} \cdot \underset{1}{\cancel{9}} \cdot 3}$

$= \dfrac{5}{12}$

d. $\dfrac{5}{6} \cdot \dfrac{1}{3} \cdot \dfrac{18}{25} = \dfrac{5 \cdot 1 \cdot 18}{6 \cdot 3 \cdot 25}$

$= \dfrac{\overset{1}{\cancel{5}} \cdot 1 \cdot \overset{1}{\cancel{6}} \cdot \overset{1}{\cancel{3}}}{\underset{1}{\cancel{6}} \cdot \underset{1}{\cancel{3}} \cdot \underset{1}{\cancel{5}} \cdot 5}$

$= \dfrac{1}{5}$

e. $\dfrac{3}{5} \cdot 7 = \dfrac{3}{5} \cdot \dfrac{7}{1}$

$= \dfrac{3 \cdot 7}{5 \cdot 1}$

$= \dfrac{21}{5}$

11. f. $(-4)\left(-\dfrac{9}{16}\right) = \left(-\dfrac{4}{1}\right)\left(-\dfrac{9}{16}\right)$

$= \dfrac{4 \cdot 9}{1 \cdot 16}$

$= \dfrac{\overset{1}{\cancel{4}} \cdot 9}{1 \cdot 4 \cdot \underset{1}{\cancel{4}}}$

$= \dfrac{9}{4}$

g. $3\left(\dfrac{1}{3}\right) = \dfrac{3}{1}\left(\dfrac{1}{3}\right)$

$= \dfrac{3 \cdot 1}{1 \cdot 3}$

$= \dfrac{\overset{1}{\cancel{3}} \cdot 1}{1 \cdot \underset{1}{\cancel{3}}}$

$= \dfrac{1}{1}$

$= 1$

h. $-\dfrac{6}{7}\left(-\dfrac{7}{6}\right) = \dfrac{6 \cdot 7}{7 \cdot 6}$

$= \dfrac{\overset{1}{\cancel{6}} \cdot \overset{1}{\cancel{7}}}{\underset{1}{\cancel{7}} \cdot \underset{1}{\cancel{6}}}$

$= \dfrac{1}{1}$

$= 1$

13. a. $\dfrac{3t}{5} \cdot \dfrac{10}{27t} = \dfrac{3t \cdot 10}{5 \cdot 27t}$

$= \dfrac{\overset{1}{\cancel{3}} \cdot \overset{1}{\cancel{t}} \cdot \overset{1}{\cancel{5}} \cdot 2}{\underset{1}{\cancel{5}} \cdot 9 \cdot \underset{1}{\cancel{3}} \cdot \underset{1}{\cancel{t}}}$

$= \dfrac{2}{9}$

13. b. $-\dfrac{2}{3}\left(\dfrac{4}{7}s\right) = -\dfrac{2 \cdot 4 \cdot s}{3 \cdot 7}$

$$= -\dfrac{8s}{21}$$

$$= -\dfrac{8}{21}s$$

c. $\dfrac{4d^2}{9x} \cdot \dfrac{3x}{28d} = \dfrac{4d^2 \cdot 3x}{9x \cdot 28d}$

$$= \dfrac{\overset{1}{4} \cdot \overset{1}{d} \cdot d \cdot \overset{1}{3} \cdot \overset{1}{x}}{3 \cdot \underset{1}{3} \cdot \underset{1}{x} \cdot 7 \cdot \underset{1}{4} \cdot \underset{1}{d}}$$

$$= \dfrac{d}{21}$$

d. $9mn\left(-\dfrac{5}{81n^2}\right) = \dfrac{9mn}{1}\left(-\dfrac{5}{81n^2}\right)$

$$= -\dfrac{9mn \cdot 5}{1 \cdot 81n^2}$$

$$= -\dfrac{\overset{1}{9} \cdot m \cdot \overset{1}{n} \cdot 5}{1 \cdot \underset{1}{9} \cdot 9 \cdot \underset{1}{n} \cdot n}$$

$$= -\dfrac{5m}{9n}$$

15. $\dfrac{1}{6}(180) = \dfrac{1}{6}\left(\dfrac{180}{1}\right)$

$$= \dfrac{1 \cdot 180}{6 \cdot 1}$$

$$= \dfrac{1 \cdot 30 \cdot \overset{1}{6}}{\underset{1}{6} \cdot 1}$$

$$= \dfrac{30}{1}$$

$$= 30$$

The astronaut will weigh 30 lb on the moon.

17. a. The reciprocal of $\dfrac{1}{8}$ is 8.

b. The reciprocal of $-\dfrac{11}{12}$ is $-\dfrac{12}{11}$.

c. The reciprocal of x is $\dfrac{1}{x}$.

d. The reciprocal of $\dfrac{ab}{c}$ is $\dfrac{c}{ab}$.

19. a. $\dfrac{t}{8} \div \dfrac{1}{4} = \dfrac{t}{8} \cdot \dfrac{4}{1}$

$$= \dfrac{t \cdot 4}{8 \cdot 1}$$

$$= \dfrac{t \cdot \overset{1}{4}}{4 \cdot 2 \cdot 1}$$

$$= \dfrac{t}{2}$$

b. $\dfrac{4a}{5} \div \dfrac{a}{2} = \dfrac{4a}{5} \cdot \dfrac{2}{a}$

$$= \dfrac{4a \cdot 2}{5 \cdot a}$$

$$= \dfrac{4 \cdot \overset{1}{a} \cdot 2}{5 \cdot \underset{1}{a}}$$

$$= \dfrac{8}{5}$$

19. c.

$$-\frac{a}{b} \div \left(-\frac{b}{a}\right) = -\frac{a}{b}\left(-\frac{a}{b}\right)$$

$$= \frac{a \cdot a}{b \cdot b}$$

$$= \frac{a^2}{b^2}$$

d.

$$\frac{2}{3}x \div \left(-\frac{x^2}{9}\right) = \frac{2x}{3}\left(-\frac{9}{x^2}\right)$$

$$= -\frac{2x \cdot 9}{3 \cdot x^2}$$

$$= -\frac{2 \cdot \overset{1}{\cancel{x}} \cdot \overset{1}{\cancel{3}} \cdot 3}{\underset{1}{\cancel{3}} \cdot \underset{1}{\cancel{x}} \cdot x}$$

$$= -\frac{6}{x}$$

21. a.

$$\frac{2}{7} + \frac{3}{7} = \frac{2+3}{7}$$

$$= \frac{5}{7}$$

b.

$$-\frac{3}{5} - \frac{3}{5} = \frac{-3-3}{5}$$

$$= -\frac{6}{5}$$

c.

$$\frac{3}{x} - \frac{1}{x} = \frac{3-1}{x}$$

$$= \frac{2}{x}$$

d.

$$\frac{7}{8} + \frac{t}{8} = \frac{7+t}{8}$$

23. $45 = 5 \cdot 3 \cdot 3$
$30 = 5 \cdot 3 \cdot 2$
$\text{LCM} = 5 \cdot 3 \cdot 3 \cdot 2 = 90$

25.

$$\frac{3}{4} - \frac{17}{32} = \frac{3 \cdot 8}{4 \cdot 8} - \frac{17}{32}$$

$$= \frac{24}{32} - \frac{17}{32}$$

$$= \frac{7}{32}$$

$\dfrac{7}{32}$ in. must be milled off.

27. $2\dfrac{1}{6} = \dfrac{13}{6}$

29. a.

$$9\frac{3}{8} = \frac{9(8)+3}{8}$$

$$= \frac{72+3}{8}$$

$$= \frac{75}{8}$$

b.

$$-2\frac{1}{5} = -\frac{2(5)+1}{5}$$

$$= -\frac{10+1}{5}$$

$$= -\frac{11}{5}$$

c.

$$100\frac{1}{2} = \frac{100(2)+1}{2}$$

$$= \frac{200+1}{2}$$

$$= \frac{201}{2}$$

d.

$$1\frac{99}{100} = \frac{100(1)+99}{100}$$

$$= \frac{100+99}{100}$$

$$= \frac{199}{100}$$

31. a.

$$-5\frac{1}{4} \cdot \frac{2}{35} = -\frac{21}{4} \cdot \frac{2}{35}$$

$$= -\frac{21 \cdot 2}{4 \cdot 35}$$

$$= -\frac{\overset{1}{\cancel{7}} \cdot 3 \cdot \overset{1}{\cancel{2}}}{2 \cdot \underset{1}{\cancel{2}} \cdot \underset{1}{\cancel{7}} \cdot 5}$$

$$= -\frac{3}{10}$$

b.

$$\left(-3\frac{1}{2}\right) \div \left(-3\frac{2}{3}\right) = \left(-\frac{7}{2}\right) \div \left(-\frac{11}{3}\right)$$

$$= -\frac{7}{2}\left(-\frac{3}{11}\right)$$

$$= \frac{7 \cdot 3}{2 \cdot 11}$$

$$= \frac{21}{22}$$

c.

$$\left(-6\frac{2}{3}\right)(-6) = -\frac{20}{3}\left(-\frac{6}{1}\right)$$

$$= \frac{20 \cdot 6}{3 \cdot 1}$$

$$= \frac{20 \cdot \overset{1}{\cancel{3}} \cdot 2}{\underset{1}{\cancel{3}} \cdot 1}$$

$$= \frac{40}{1}$$

$$= 40$$

31. d.

$$-8 \div 3\frac{1}{5} = -\frac{8}{1} \div \frac{16}{5}$$

$$= -\frac{8}{1} \cdot \frac{5}{16}$$

$$= -\frac{8 \cdot 5}{1 \cdot 16}$$

$$= -\frac{\overset{1}{\cancel{8}} \cdot 5}{1 \cdot \underset{1}{\cancel{8}} \cdot 2}$$

$$= -\frac{5}{2}$$

$$= -2\frac{1}{2}$$

33. a.
$$1\frac{3}{8} + 2\frac{1}{5} = \frac{11}{8} + \frac{11}{5}$$
$$= \frac{11 \cdot 5}{8 \cdot 5} + \frac{11 \cdot 8}{5 \cdot 8}$$
$$= \frac{55}{40} + \frac{88}{40}$$
$$= \frac{143}{40}$$
$$= 3\frac{23}{40}$$

33. c.
$$2\frac{5}{6} - 1\frac{3}{4} = \frac{17}{6} - \frac{7}{4}$$
$$= \frac{17 \cdot 2}{6 \cdot 2} - \frac{7 \cdot 3}{4 \cdot 3}$$
$$= \frac{34}{12} - \frac{21}{12}$$
$$= \frac{13}{12}$$
$$= 1\frac{1}{12}$$

b.
$$3\frac{1}{2} + 2\frac{2}{3} = \frac{7}{2} + \frac{8}{3}$$
$$= \frac{7 \cdot 3}{2 \cdot 3} + \frac{8 \cdot 2}{3 \cdot 2}$$
$$= \frac{21}{6} + \frac{16}{6}$$
$$= \frac{37}{6}$$
$$= 6\frac{1}{6}$$

d.
$$3\frac{7}{16} - 2\frac{1}{8} = \frac{55}{16} - \frac{17}{8}$$
$$= \frac{55}{16} - \frac{17 \cdot 2}{8 \cdot 2}$$
$$= \frac{55}{16} - \frac{34}{16}$$
$$= \frac{21}{16}$$
$$= 1\frac{5}{16}$$

35. a.
$$33\frac{1}{9} \quad = \quad 33\frac{1 \cdot 2}{9 \cdot 2} \quad = \quad 33\frac{2}{18}$$
$$+49\frac{1}{6} \quad = +49\frac{1 \cdot 3}{6 \cdot 3} \quad = +49\frac{3}{18}$$
$$\overline{\qquad\qquad\qquad\qquad\qquad 82\frac{5}{18}}$$

35. b.

$$98\frac{11}{20} \quad = \quad 98\frac{11}{20} \quad = \quad 98\frac{11}{20}$$

$$+14\frac{3}{5} \quad = \quad +14\frac{3\cdot4}{5\cdot4} \quad = \quad +14\frac{12}{20}$$

$$112\frac{23}{20}$$

$$112\frac{23}{20} = 112 + 1\frac{3}{20} = 113\frac{3}{20}$$

c.

$$50\frac{5}{8} \quad = \quad 50\frac{5\cdot3}{8\cdot3} \quad = \quad 50\frac{15}{24}$$

$$-19\frac{1}{6} \quad = \quad -19\frac{1\cdot4}{6\cdot4} \quad = \quad -19\frac{4}{24}$$

$$31\frac{11}{24}$$

d.

$$75\frac{3}{4}$$

$$-59$$

$$16\frac{3}{4}$$

37. a.

$$\frac{3}{4} + \left(-\frac{1}{3}\right)^2\left(\frac{5}{4}\right) = \frac{3}{4} + \frac{1}{9}\left(\frac{5}{4}\right)$$

$$= \frac{3}{4} + \frac{5}{36}$$

$$= \frac{3\cdot9}{4\cdot9} + \frac{5}{36}$$

$$= \frac{27}{36} + \frac{5}{36}$$

$$= \frac{32}{36}$$

$$= \frac{8\cdot\overset{1}{4}}{9\cdot\underset{1}{4}}$$

$$= \frac{8}{9}$$

37. b.
$$\left(\frac{2}{3} \div \frac{16}{9}\right) - \left(1\frac{2}{3} \cdot \frac{1}{15}\right) = \left(\frac{2}{3} \cdot \frac{9}{16}\right) - \left(\frac{5}{3} \cdot \frac{1}{15}\right)$$

$$= \left(\frac{2 \cdot 9}{3 \cdot 16}\right) - \left(\frac{5 \cdot 1}{3 \cdot 15}\right)$$

$$= \frac{\overset{1}{\cancel{2}} \cdot \overset{1}{\cancel{3}} \cdot 3}{\underset{1}{\cancel{3}} \cdot 8 \cdot \underset{1}{\cancel{2}}} - \frac{\overset{1}{\cancel{5}} \cdot 1}{3 \cdot \underset{1}{\cancel{5}} \cdot 3}$$

$$= \frac{3}{8} - \frac{1}{9}$$

$$= \frac{3 \cdot 9}{8 \cdot 9} - \frac{1 \cdot 8}{9 \cdot 8}$$

$$= \frac{27}{72} - \frac{8}{72}$$

$$= \frac{19}{72}$$

39. a.
$$d^2 - 2c = \left(\frac{1}{8}\right)^2 - 2\left(-\frac{3}{4}\right)$$

$$= \frac{1}{64} - \frac{2}{1}\left(\frac{-3}{4}\right)$$

$$= \frac{1}{64} - \left(-\frac{6}{4}\right)$$

$$= \frac{1}{64} + \frac{6 \cdot 16}{4 \cdot 16}$$

$$= \frac{1}{64} + \frac{96}{64}$$

$$= \frac{97}{64}$$

39. b.

$$-cd + e = -\left(-\frac{3}{4}\right)\left(\frac{1}{8}\right) + \left(-2\frac{1}{16}\right)$$

$$= -1\left(-\frac{3}{4}\right)\left(\frac{1}{8}\right) + \left(-\frac{33}{16}\right)$$

$$= \frac{3 \cdot 1}{4 \cdot 8} + \left(-\frac{33}{16}\right)$$

$$= \frac{3}{32} + \left(-\frac{33}{16}\right)$$

$$= \frac{3}{32} + \left(-\frac{33 \cdot 2}{16 \cdot 2}\right)$$

$$= \frac{3}{32} + \left(-\frac{66}{32}\right)$$

$$= -\frac{63}{32}$$

c.

$$e \div (cd) = -2\frac{1}{16} \div \left[\left(-\frac{3}{4}\right)\left(\frac{1}{8}\right)\right]$$

$$= -\frac{33}{16} \div \left(-\frac{3 \cdot 1}{4 \cdot 8}\right)$$

$$= -\frac{33}{16} \div \left(-\frac{3}{32}\right)$$

$$= -\frac{33}{16}\left(-\frac{32}{3}\right)$$

$$= \frac{33 \cdot 32}{16 \cdot 3}$$

$$= \frac{11 \cdot \overset{1}{\cancel{3}} \cdot \overset{1}{\cancel{16}} \cdot 2}{\cancel{16} \cdot \cancel{3}}$$

$$= \frac{22}{1}$$

$$= 22$$

39. d.

$$\frac{c - d}{e} = \frac{-\dfrac{3}{4} - \dfrac{1}{8}}{-2\dfrac{1}{16}}$$

$$= \frac{-\dfrac{3}{4} - \dfrac{1}{8}}{-\dfrac{33}{16}}$$

$$= \frac{16\left(-\dfrac{3}{4} - \dfrac{1}{8}\right)}{16\left(-\dfrac{33}{16}\right)}$$

$$= \frac{-12 - 2}{-33}$$

$$= \frac{-14}{-33}$$

$$= \frac{14}{33}$$

41. a.

$$\frac{c}{3} - \frac{c}{8} = 2$$

$$24\left(\frac{c}{3} - \frac{c}{8}\right) = 24(2)$$

$$8c - 3c = 48$$

$$5c = 48$$

$$\frac{5c}{5} = \frac{48}{5}$$

$$c = \frac{48}{5}$$

b.

$$\frac{5h}{9} - 1 = -3$$

$$9\left(\frac{5h}{9} - 1\right) = 9(-3)$$

$$5h - 9 = -27$$

$$5h - 9 + 9 = -27 + 9$$

$$5h = -18$$

$$\frac{5h}{5} = \frac{-18}{5}$$

$$h = -\frac{18}{5}$$

c.

$$4 - \frac{d}{4} = 0$$

$$4 - \frac{d}{4} + \frac{d}{4} = 0 + \frac{d}{4}$$

$$4 = \frac{d}{4}$$

$$4(4) = 4\left(\frac{d}{4}\right)$$

$$16 = d$$

41. d.

$$\frac{t}{10} - \frac{2}{3} = \frac{1}{5}$$

$$30\left(\frac{t}{10} - \frac{2}{3}\right) = 30\left(\frac{1}{5}\right)$$

$$3t - 20 = 6$$

$$3t - 20 + 20 = 6 + 20$$

$$3t = 26$$

$$\frac{3t}{3} = \frac{26}{3}$$

$$t = \frac{26}{3}$$

CHAPTER 4 TEST

1. a. $\frac{4}{5}$ of the plant is above the ground.

b. $\frac{1}{5}$ of the plant is below ground.

3.

$$-\frac{3x}{4}\left(\frac{1}{5x^2}\right) = -\frac{3x \cdot 1}{4 \cdot 5x^2}$$

$$= -\frac{3 \cdot \overset{1}{\cancel{x}} \cdot 1}{4 \cdot 5 \cdot \cancel{x} \cdot x}$$

$$= -\frac{3}{20x}$$

5. $\dfrac{4ab}{3} \div \dfrac{a^2b}{9} = \dfrac{4ab}{3} \cdot \dfrac{9}{a^2b}$

$\qquad = \dfrac{4ab \cdot 9}{3 \cdot a^2b}$

$\qquad = \dfrac{4 \cdot \overset{1}{\cancel{a}} \cdot \overset{1}{\cancel{b}} \cdot \overset{1}{\cancel{3}} \cdot 3}{\underset{1}{\cancel{3}} \cdot \underset{1}{\cancel{a}} \cdot a \cdot \underset{1}{\cancel{b}}}$

$\qquad = \dfrac{12}{a}$

7. $\dfrac{7}{8} = \dfrac{7 \cdot 3a}{8 \cdot 3a}$

$\qquad = \dfrac{21a}{24a}$

9. $13\dfrac{1}{2} \div 9 = \dfrac{27}{2} \div \dfrac{9}{1}$

$\qquad = \dfrac{27}{2} \cdot \dfrac{1}{9}$

$\qquad = \dfrac{27 \cdot 1}{2 \cdot 9}$

$\qquad = \dfrac{\overset{1}{\cancel{9}} \cdot 3 \cdot 1}{2 \cdot \underset{1}{\cancel{9}}}$

$\qquad = \dfrac{3}{2}$

$\qquad = 1\dfrac{1}{2}$

The player earns $\$1\dfrac{1}{2}$ million per year.

11. $57\dfrac{5}{9} + 103\dfrac{3}{4} = 57 + \dfrac{5}{9} + 103 + \dfrac{3}{4}$

$\qquad = 57 + 103 + \dfrac{5}{9} + \dfrac{3}{4}$

$\qquad = 160 + \dfrac{5 \cdot 4}{9 \cdot 4} + \dfrac{3 \cdot 9}{4 \cdot 9}$

$\qquad = 160 + \dfrac{20}{36} + \dfrac{27}{36}$

$\qquad = 160 + \dfrac{47}{36}$

$\qquad = 160 + 1\dfrac{11}{36}$

$\qquad = 161\dfrac{11}{36}$

13. $-\dfrac{3}{7} + 2 = -\dfrac{3}{7} + \dfrac{2}{1}$

$\qquad = -\dfrac{3}{7} + \dfrac{2 \cdot 7}{1 \cdot 7}$

$\qquad = -\dfrac{3}{7} + \dfrac{14}{7}$

$\qquad = \dfrac{-3 + 14}{7}$

$\qquad = \dfrac{11}{7}$

15.
$$P = a + b + c$$
$$P = 8 + 8\frac{3}{4} + 3\frac{1}{2}$$
$$= 8 + 8 + \frac{3}{4} + 3 + \frac{1}{2}$$
$$= 8 + 8 + 3 + \frac{3}{4} + \frac{1}{2}$$
$$= 19 + \frac{3}{4} + \frac{1 \cdot 2}{2 \cdot 2}$$
$$= 19 + \frac{3}{4} + \frac{2}{4}$$
$$= 19 + \frac{5}{4}$$
$$= 19 + 1\frac{1}{4}$$
$$= 20\frac{1}{4}$$

$$A = \frac{1}{2}bh$$
$$A = \frac{1}{2}\left(3\frac{1}{2}\right)(8)$$
$$= \frac{1}{2}\left(\frac{7}{2}\right)\left(\frac{8}{1}\right)$$
$$= \frac{1 \cdot 7 \cdot 8}{2 \cdot 2 \cdot 1}$$
$$= \frac{1 \cdot 7 \cdot \overset{1}{2} \cdot 2 \cdot \overset{1}{2}}{2 \cdot 2 \cdot 1}$$
$$= \frac{14}{1}$$
$$= 14$$

The perimeter is $20\frac{1}{4}$ in. and the area is 14 in.^{2}.

17.
$$\frac{-\frac{5}{6}}{\frac{7}{8}} = -\frac{5}{6} \div \frac{7}{8}$$
$$= -\frac{5}{6} \cdot \frac{8}{7}$$
$$= -\frac{5 \cdot 8}{6 \cdot 7}$$
$$= -\frac{5 \cdot 4 \cdot \overset{1}{2}}{3 \cdot 2 \cdot 7}$$
$$= -\frac{20}{21}$$

19. a.
$$\frac{x}{3} = 14$$
$$3\left(\frac{x}{3}\right) = 3(14)$$
$$x = 42$$

b.
$$-\frac{5}{2}t = 18$$
$$-\frac{2}{5}\left(-\frac{5}{2}t\right) = -\frac{2}{5}(18)$$
$$t = -\frac{36}{5}$$

21. a. Dividing the numerator and denominator of a fraction by the same number

b. Equivalent fractions: $\dfrac{1}{2} = \dfrac{2}{4}$

c. Multiplying the numerator and denominator of a fraction by the same number

1. 5,434,700

3.
$$\begin{array}{r} 4,679 \\ +3,457 \\ \hline 8,136 \end{array}$$

5.
$$\begin{array}{r} 5,345 \\ \times \quad 56 \\ \hline 32070 \\ 26725 \\ \hline 299,320 \end{array}$$

7. $P = 2l + 2w$
$P = 2(150) + 2(75)$
$\quad = 300 + 150$
$\quad = 450$
The perimeter is 450 ft.

9. $84 = 2^2 \cdot 3 \cdot 7$

11. $360 = 2^3 \cdot 3^2 \cdot 5$

13. $6 + (-2)(-5) = 6 + 10$
$\quad\quad\quad\quad = 16$

15. $\dfrac{2(-7) + 3(2)}{2(-2)} = \dfrac{-14 + 6}{-4}$
$\quad\quad\quad\quad = \dfrac{-8}{-4}$
$\quad\quad\quad\quad = 2$

17. $x + 15$

19. $4x$

21. $2x - 1 = 2(4) - 1$
$\quad\quad\quad = 8 - 1$
$\quad\quad\quad = 7$

23. $3x - x^3 = 3(4) - 4^3$
$\quad\quad\quad\quad = 12 - 64$
$\quad\quad\quad\quad = -52$

25. $-3(5x) = -15x$

27. $-2(3x - 4) = -6x + 8$

29. $-3x + 8x = 5x$

31. $4x - 3y - 5x + 2y = -x - y$

33.
$$3x + 2 = -13$$
$$3x + 2 - 2 = -13 - 2$$
$$3x = -15$$
$$\frac{3x}{3} = \frac{-15}{3}$$
$$x = -5$$

35.
$$\frac{y}{4} - 1 = -5$$
$$\frac{y}{4} - 1 + 1 = -5 + 1$$
$$\frac{y}{4} = -4$$
$$4\left(\frac{y}{4}\right) = 4(-4)$$
$$y = -16$$

37.
$$6x - 12 = 2x + 4$$
$$6x - 12 - 2x = 2x + 4 - 2x$$
$$4x - 12 = 4$$
$$4x - 12 + 12 = 4 + 12$$
$$4x = 16$$
$$\frac{4x}{4} = \frac{16}{4}$$
$$x = 4$$

39. Let x = the number of 3-hour shifts needed.

$$37 + 3x = 100$$
$$37 + 3x - 37 = 100 - 37$$
$$3x = 63$$
$$\frac{3x}{3} = \frac{63}{3}$$
$$x = 21$$

The student needs 21 more 3-hour shifts.

41. $p^3 p^5 = p^{3+5}$
$$= p^8$$

43. $(-x^2 y^3)(x^3 y^4) = -x^{2+3} y^{3+4}$
$$= -x^5 y^7$$

45. $(-n^2)^3 = -n^{2 \cdot 3}$
$$= -n^6$$

47. $(2p^3)^2 (3p^2)^3 = (2^2 \cdot p^{3 \cdot 2})(3^3 p^{2 \cdot 3})$
$$= 4p^6 \cdot 27 p^6$$
$$= 4 \cdot 27 \cdot p^{6+6}$$
$$= 108 p^{12}$$

49. $\dfrac{21}{28} = \dfrac{\overset{1}{\cancel{7}} \cdot 3}{\underset{1}{\cancel{7}} \cdot 4}$

$$= \frac{3}{4}$$

50. $\dfrac{40x^6 y^4}{16x^3 y^5} = \dfrac{8 \cdot 5 \cdot \overset{1}{\cancel{x}} \cdot \overset{1}{\cancel{x}} \cdot \overset{1}{\cancel{x}} \cdot x \cdot x \cdot x \cdot \overset{1}{\cancel{y}} \cdot \overset{1}{\cancel{y}} \cdot \overset{1}{\cancel{y}} \cdot \overset{1}{\cancel{y}} \cdot y}{8 \cdot 2 \cdot \underset{1}{\cancel{x}} \cdot \underset{1}{\cancel{x}} \cdot \underset{1}{\cancel{x}} \cdot \underset{1}{\cancel{y}} \cdot \underset{1}{\cancel{y}} \cdot \underset{1}{\cancel{y}} \cdot \underset{1}{\cancel{y}} \cdot y}$

$$= \frac{5x^3}{2y}$$

51. $\dfrac{6}{5} \cdot \dfrac{2}{3} = \dfrac{6 \cdot 2}{5 \cdot 3}$

$= \dfrac{\overset{1}{\cancel{3}} \cdot 2 \cdot 2}{5 \cdot \underset{1}{\cancel{3}}}$

$= \dfrac{4}{5}$

53. $\dfrac{2}{3} + \dfrac{3}{4} = \dfrac{2 \cdot 4}{3 \cdot 4} + \dfrac{3 \cdot 3}{4 \cdot 3}$

$= \dfrac{8}{12} + \dfrac{9}{12}$

$= \dfrac{17}{12}$

$= 1\dfrac{5}{12}$

55. $3\dfrac{5}{6} = \dfrac{3(6) + 5}{6}$

$= \dfrac{18 + 5}{6}$

$= \dfrac{23}{6}$

57. $4\dfrac{2}{3} + 5\dfrac{1}{4} = \dfrac{14}{3} + \dfrac{21}{4}$

$= \dfrac{14 \cdot 4}{3 \cdot 4} + \dfrac{21 \cdot 3}{4 \cdot 3}$

$= \dfrac{56}{12} + \dfrac{63}{12}$

$= \dfrac{119}{12}$

$= 9\dfrac{11}{12}$

59. $\left(\dfrac{1}{4} - \dfrac{7}{8}\right) \div \left(-2\dfrac{3}{16}\right) = \left(\dfrac{1 \cdot 2}{4 \cdot 2} - \dfrac{7}{8}\right) \div \left(-\dfrac{35}{16}\right)$

$= \left(\dfrac{2}{8} - \dfrac{7}{8}\right) \div \left(-\dfrac{35}{16}\right)$

$= -\dfrac{5}{8} \div \left(-\dfrac{35}{16}\right)$

$= -\dfrac{5}{8}\left(-\dfrac{16}{35}\right)$

$= \dfrac{5 \cdot 16}{8 \cdot 35}$

$= \dfrac{\overset{1}{\cancel{5}} \cdot \overset{1}{\cancel{8}} \cdot 2}{\underset{1}{\cancel{8}} \cdot 7 \cdot \underset{1}{\cancel{5}}}$

$= \dfrac{2}{7}$

61.

$$x + \frac{1}{5} = -\frac{14}{15}$$

$$x + \frac{1}{5} - \frac{1}{5} = -\frac{14}{15} - \frac{1}{5}$$

$$x = -\frac{14}{15} - \frac{1 \cdot 3}{5 \cdot 3}$$

$$x = -\frac{14}{15} - \frac{3}{15}$$

$$x = -\frac{17}{15}$$

STUDY SET Section 5.1

VOCABULARY

1. tens; ones; tenths; hundredths; thousandths; ten thousandths

3. We can approximate a decimal number using the process called <u>rounding</u>.

CONCEPTS

5. a. thirty-two and four hundred fifteen thousandths

   b. 32 is its whole number part.

   c. $\dfrac{415}{1,000}$ is its fractional part.

   d. $32.415 = 30 + 2 + \dfrac{4}{10} + \dfrac{1}{100} + \dfrac{5}{1,000}$

7.

9. a. $0.9 = 0.90$ is true

   b. $1.260 = 1.206$ is false

   c. $-1.2800 = -1.280$ is true

   d. $0.001 = .0010$ is true

11. $\dfrac{47}{100}$, 0.47

13.

NOTATION

15. 9,816.0245

PRACTICE

17. fifty and one tenth; $50\dfrac{1}{10}$

19. negative one hundred thirty-seven ten thousandths; $-\dfrac{137}{10,000}$

21. three hundred four and three ten thousandths; $304\dfrac{3}{10,000}$

23. negative seventy-two and four hundred ninety-three thousandths; $-72\dfrac{493}{1,000}$

25. -0.39

27. 6.187

29. 506.1

31. 77.2

33. -0.14

35. 33.00

37. 3.233

39. 55.039

41. 39

43. 2,988

45. a. $3,090

   b. $3,090.30

47. −23.45 ≤ −23.1

49. −.065 ≥ −.066

51. 132.64, 132.6401, 132.6499

APPLICATIONS

53.

Sample	Location	Size (in.)	Classification
A	NE corner	0.095	**granule**
B	riverbank	0.009	**sand**
C	hilltop	0.0007	**silt**
D	dry lake	0.00003	**clay**

55. The horse begins strong, slows down for the middle two splits, and finishes strong.

57. a. 0.91

   b. 0.30

   c. 1,609.34

   d. 453.59

   e. 28.35

   f. 3.79

59.

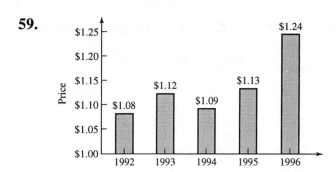

61.

Dollars	Cents
$0.50	**50**
$0.05	**5**
$0.55	**55**
$5.00	**500**
$0.01	**1**

REVIEW

67.

$$75\frac{3}{4} \;=\; 75\frac{3\cdot5}{4\cdot5} \;=\; 75\frac{15}{20}$$

$$+\;88\frac{4}{5} \;=\; +\;88\frac{4\cdot4}{5\cdot4} \;=\; +\;88\frac{16}{20}$$

$$163\frac{31}{20}$$

$$163\frac{31}{20} = 163 + \frac{31}{20}$$

$$= 163 + 1\frac{11}{20}$$

$$= 164\frac{11}{20}$$

69.
$$5R - 3(6 - R) = 5R - 3(6) - (-3)R$$
$$= 5R - 18 - (-3R)$$
$$= 5R - 18 + 3R$$
$$= 8R - 18$$

71.
$$A = \frac{1}{2}bh$$
$$A = \frac{1}{2}(16)(9)$$
$$= 8(9)$$
$$= 72$$

The area of the triangle is 72 in.^{2}

73.
$$-2 + (-3) + 4 = -5 + 4$$
$$= -1$$

STUDY SET Section 5.2

VOCABULARY

1. The answer to an addition problem is called the <u>sum</u>.

3. Every whole number has an <u>unwritten</u> sign.

CONCEPTS

5. a. $0.3 + 0.17 = 0.47$

b. $0.3 = \dfrac{3}{10}$; $0.17 = \dfrac{17}{100}$

c.
$$\frac{3}{10} + \frac{17}{100} = \frac{3 \cdot 10}{10 \cdot 10} + \frac{17}{100}$$
$$= \frac{30}{100} + \frac{17}{100}$$
$$= \frac{47}{100}$$

d. $\dfrac{47}{100} = 0.47$

5. e. The answers are the same.

PRACTICE

7.
```
  32.5
+  7.4
------
  39.9
```

9.
```
  21.6
+33.12
------
 54.72
```

11. $12 + 3.9 = 15.9$

13. $0.03034 + 0.2003 = 0.23064$

15.
```
  247.9
   40
+   0.56
-------
 288.46
```

17. $45 + 9.9 + 0.12 + 3.02 = 58.04$

19.
```
  12.98
-  3.45
------
   9.53
```

21.
```
  78.1
-  7.81
------
 70.29
```

23. $5 - 0.023 = 4.977$

25. $24 - 23.81 = 0.19$

27. $-45.6 + 34.7 = -10.9$

29. $46.09 + (-7.8) = 38.29$

31. $-7.8 + (-6.5) = -14.3$

33. $-0.0045 + (-0.031) = -0.0355$

35. $-9.5 - 7.1 = -16.6$

37. $30.03 - (-17.88) = 47.91$

39. $-2.002 - (-4.6) = 2.598$

41. $-7 - (-18.01) = 11.01$

43. $3.4 - 6.6 + 7.3 = (3.4 - 6.6) + 7.3$
$$= [3.4 + (-6.6)] + 7.3$$
$$= -3.2 + 7.3$$
$$= 4.1$$

45. $(-9.1 - 6.05) - (-51) = [-9.1 + (-6.05)] - (-51)$
$$= -15.15 + [-(-51)]$$
$$= -15.15 + 51$$
$$= 35.85$$

47. $16 - (67.2 + 6.27) = 16 - 73.47$
$$= 16 + (-73.47)$$
$$= -57.47$$

49. $(-7.2 + 6.3) - (-3.1 - 4) = -0.9 - [-3.1 + (-4)]$
$$= -0.9 - (-7.1)$$
$$= -0.9 + [-(-7.1)]$$
$$= -0.9 + 7.1$$
$$= 6.2$$

51. $2.43 + 5.6 = 8.03$

APPLICATIONS

53. a. $53.03 + 0.014 = 53.044$
The Italian bobsled team's time was 53.044 sec.

b. $102.71 - 0.33 = 102.38$
The second place finisher's total was 102.38.

55. $187.8 - (43.5 + 40.9) = 187.8 - 84.4$
$$= 103.4$$
The wheel base is 103.4 in.

57. $30.7 - 28.9 = 30.7 + (-28.9)$
$$= 1.8$$
The difference between the highest and lowest pressure is 1.8.
You would expect the weather to be fair in Texas.

59.

	Pipe underwater	Pipe underground	Total pipe
Design 1	1.74 mi	2.32 mi	4.06 mi
Design 2	2.90 mi	0 mi	2.90 mi

61. $54.48 - 10.49 = 43.99$
She ran 43.99 sec faster than Thompson swam.

63. subtotal $= 242.50 + 116.10 + 47.93 + 359.16$
$$= 765.69$$
The subtotal is $765.69.

65. 1994: $3.6 - 1.7 = 1.6$
They had a profit of $1.9 million.
1995: $4.8 - 5.2 = -0.4$
They had a loss of $.4 million.
1996 : $4.4 - 2.7 = 1.7$
They had a profit of $1.7 million.

67. $2,367.909 + 5,789.0253 = 8,156.9343$

69. $9,000.09 - 7,067.445 = 1,932.645$

71. $3,434.768 - (908 - 2.3 + .0098) = 3,434.768 - 905.7098$
$$= 2,529.0582$$

REVIEW

75. $\left(-2x^3\right)^3 = (-2)^3 (x^3)^3$
$$= -8x^{3 \cdot 3}$$
$$= -8x^9$$

77. $\dfrac{-15}{26} \cdot 1\dfrac{4}{9} = \dfrac{-15}{26} \cdot \dfrac{13}{9}$
$$= -\dfrac{15 \cdot 13}{26 \cdot 9}$$
$$= -\dfrac{5 \cdot \overset{1}{\cancel{3}} \cdot \overset{1}{\cancel{13}}}{\cancel{13} \cdot 2 \cdot \cancel{3} \cdot 3}$$
$$= -\dfrac{5}{6}$$

STUDY SET Section 5.3

VOCABULARY

1. In the multiplication problem 2.89 • 15.7, 2.89 and 15.7 are called <u>factors</u>. The answer, 45.373, is called the <u>product</u>.

CONCEPTS

3. To multiply decimals, multiply them as if they were <u>whole</u> numbers. The number of decimal places in the product is the same as the <u>sum</u> of the decimal places of the factors.

5. When we move the decimal point to the right, the decimal number gets larger.

7. a. $\dfrac{3}{10} \cdot \dfrac{7}{100} = \dfrac{21}{1,000}$

b. $0.3(0.07) = 0.021$

$0.021 = \dfrac{21}{1000}.$

They are the same.

PRACTICE

9. $(0.4)(0.2) = 0.08$

11. $(-0.5)(0.3) = -0.15$

13. $(1.4)(0.7) = 0.98$

15. $(0.08)(0.9) = 0.072$

17. $(-5.6)(-2.2) = 12.32$

19. $(-4.9)(0.001) = -0.0049$

21. $(-0.35)(0.24) = -0.084$

23. $(-2.13)(4.05) = -8.6265$

25. $16 \cdot 0.6 = 9.6$

27. $-7(8.1) = -56.7$

29. $0.04(306) = 12.24$

31. $60.61(-0.3) = -18.183$

33. $-0.2(0.3)(-0.4) = 0.024$

35. $5.5(10)(-0.3) = -16.5$

37. $4.2 \cdot 10 = 42$

39. $67.164 \cdot 100 = 6,716.4$

41. $-0.056(10) = -0.56$

43. $1,000(8.05) = 8,050$

45. $0.098(10,000) = 980$

47. $-0.2 \cdot 1,000 = -200$

49.

Decimal	Its square
0.1	**0.01**
0.2	**0.04**
0.3	**0.09**
0.4	**0.16**
0.5	**0.25**
0.6	**0.36**
0.7	**0.49**
0.8	**0.64**
0.9	**0.81**

51. $(1.2)^2 = 1.2 \cdot 1.2$
$= 1.44$

53. $(-1.3)^2 = (-1.3)(-1.3)$
$= 1.69$

55. $-4.6(23.4 - 19.6) = -4.6(3.8)$
$= -17.48$

57. $(-0.2)^2 + 2(7.1) = 0.04 + 2(7.1)$
$= 0.04 + 14.2$
$= 14.24$

59. $(-0.7 - 0.5)(2.4 - 3.1) = (-1.2)(-0.7)$
$$= 0.84$$

61. $(0.5 + 0.6)^2(-3.2) = (1.1)^2(-3.2)$
$$= (1.21)(-3.2)$$
$$= -3.872$$

63. $3.14 + 2(d - t) = 3.14 + 2[1.2 - (-6.7)]$
$$= 3.14 + 2(7.9)$$
$$= 3.14 + 15.8$$
$$= 18.94$$

65. $t + 0.5rt^2 = -0.4 + 0.5(100)(-0.4)^2$
$$= -0.4 + 0.5(100)(.16)$$
$$= -0.4 + 50(.16)$$
$$= -0.4 + 8$$
$$= 7.6$$

APPLICATIONS

67. a.

Ticket type	Price	Number sold	Income
Floor	**$12.50**	1,000	**$12,500**
Balcony	**$15.75**	100	**$1,575**

   b. Total $= 12,500 + 1,575$
$$= 14,075$$
The total receipts were $14,075.

69. $0.57 + 2(0.09) = 0.57 + .18$
$$= 0.75$$
In the three-week period, the house fell 0.75 inches.

71. $2(45.5 + 20.5 + 2.2) = 2(68.2)$
$$= 136.4$$
The plate weight is 136.4 lb.

73.

Type of nut	Price per pound	Pounds	Cost
Almonds	$3.25	16	**$52.00**
Walnuts	$2.10	25	**$52.50**
Peanuts	$1.85	x	**$1.85x**

75. $p = 2l + 2w$

$\quad\quad = 2(50) + 2(30.3)$

$\quad\quad = 100 + 60.6$

$\quad\quad = 160.6$

The pool will need 160.6 m of coping.

77. $(-9.0089 + 10.0087)(15.3) = 15.29694$

79. $(18.18 + 6.61)^2 + (5 - 9.09)^2 = 631.2722$

81. $719(0.14277) = 102.65163$
The bill would be $102.65.

REVIEW

87.

$$\frac{x}{2} - \frac{x}{3} = -2$$

$$6\left(\frac{x}{2} - \frac{x}{3}\right) = 6(-2)$$

$$6\left(\frac{x}{2}\right) - 6\left(\frac{x}{3}\right) = 6(-2)$$

$$3x - 2x = -12$$

$$x = -12$$

89.

$$3\frac{1}{3}\left(-1\frac{4}{5}\right) = \frac{10}{3} \bullet -\frac{9}{5}$$

$$= -\frac{10 \bullet 9}{3 \bullet 5}$$

$$= -\frac{2 \bullet \overset{1}{\cancel{5}} \bullet \overset{1}{\cancel{3}} \bullet 3}{\underset{1}{\cancel{3}} \bullet \underset{1}{\cancel{5}}}$$

$$= -\frac{6}{1}$$

$$= -6$$

91. $|-3|$ is the absolute value of negative three.

93. $-\dfrac{8}{8} = -1$

1. $978.88 - 739.99 \approx 980 - 740$
≈ 240
The deluxe model is
approximately $240 more.

3. $20.6 - 18.8 \approx 21 - 19$
≈ 2
The economy model has
approximately 2 cubic feet less
storage capacity.

5. $20,000 \div 739.99 \approx 20,000 \div 700$
≈ 30
She can purchase about 30
refrigerators.

7. $978.88 \div 3 \approx 990 \div 3$
≈ 330
Each will have to pay
approximately $330.

9. $739.99 - 220 \approx 740 - 220$
≈ 520
Approximately $520 is left to
finance.

11. $53 \cdot 5.61 \approx 50 \cdot 6$
≈ 300
No, the answer does not seem
reasonable.

STUDY SET Section 5.4

VOCABULARY

1. In the division $2.5\overline{)4.075} = 1.63$,
4.075 is called the <u>dividend</u>, 2.5
is the <u>divisor</u>, and 1.63 is the
<u>quotient</u>.

CONCEPTS

3. To divide by a decimal, move the
decimal point of the divisor so
that it becomes a <u>whole</u> number.
The decimal point of the <u>dividend</u>
is then moved the same number of
places to the <u>right</u>. The decimal
point of the <u>answer</u> is written
directly above that of the
dividend.

5. $45 = 45.0 = 45.000$ is a true
statement.

7. To complete the division
$7.8\overline{)14.562}$, the decimal points of
the divisor and dividend are
moved 1 place to the right. This is
equivalent to multiplying the
numerator and the denominator of
$\dfrac{14.562}{7.8}$ by 10.

9. Use multiplication to see whether
$0.9 \cdot 2.13 = 1.917$.

NOTATION

11. The arrows show moving the
decimal points in the divisor and
dividend two places to the right.

PRACTICE

13.
$$
\begin{array}{r}
4.5 \\
8\overline{)36.0} \\
\underline{32} \\
40 \\
\underline{40} \\
0
\end{array}
$$

15.

$$4 \overline{)\begin{array}{r} 9.75 \\ 39.00 \end{array}}$$

$$\begin{array}{r} \underline{36} \\ 30 \\ \underline{28} \\ 20 \\ \underline{20} \\ 0 \end{array}$$

$$-39 \div 4 = -9.75$$

17.

$$8 \overline{)\begin{array}{r} 6.2 \\ 49.6 \end{array}}$$

$$\begin{array}{r} \underline{48} \\ 16 \\ \underline{16} \\ 0 \end{array}$$

$$49.6 \div 8 = 6.2$$

19.

$$9 \overline{)\begin{array}{r} 32.1 \\ 288.9 \end{array}}$$

$$\begin{array}{r} \underline{27} \\ 18 \\ \underline{18} \\ 09 \\ \underline{9} \\ 0 \end{array}$$

21.

$$6 \overline{)\begin{array}{r} 2.46 \\ 14.76 \end{array}}$$

$$\begin{array}{r} \underline{12} \\ 27 \\ \underline{24} \\ 36 \\ \underline{36} \\ 0 \end{array}$$

$$(-14.76) \div (-6) = 2.46$$

23.

$$7 \overline{)\begin{array}{r} 7.86 \\ 55.02 \end{array}}$$

$$\begin{array}{r} \underline{49} \\ 60 \\ \underline{56} \\ 42 \\ \underline{42} \\ 0 \end{array}$$

$$\frac{-55.02}{7} = -7.86$$

25.

$$45 \overline{)\begin{array}{r} 2.66 \\ 119.70 \end{array}}$$

$$\begin{array}{r} \underline{90} \\ 297 \\ \underline{270} \\ 270 \\ \underline{270} \\ 0 \end{array}$$

27.

$$35 \overline{)\begin{array}{r} 7.17 \\ 250.95 \end{array}}$$

$$\begin{array}{r} \underline{245} \\ 59 \\ \underline{35} \\ 245 \\ \underline{245} \\ 0 \end{array}$$

29.

$$32 \overline{)\begin{array}{r} 130 \\ 4,160 \end{array}}$$

$$\begin{array}{r} \underline{32} \\ 96 \\ \underline{96} \\ 00 \\ \underline{00} \\ 0 \end{array}$$

31.

$$
\begin{array}{r}
1{,}050 \\
19\overline{)19{,}950} \\
\underline{19}\phantom{{,}950} \\
095 \\
\underline{95} \\
00 \\
\underline{00} \\
0
\end{array}
$$

33.

$$
\begin{array}{r}
0.6 \\
17\overline{)10.2} \\
\underline{102} \\
0
\end{array}
$$

35.

$$
\begin{array}{r}
0.6 \\
31\overline{)18.6} \\
\underline{186} \\
0
\end{array}
$$

37.

$$
\begin{array}{r}
5.33 \\
3\overline{)16.00} \\
\underline{15} \\
10 \\
\underline{9} \\
10 \\
\underline{9} \\
1
\end{array}
$$

To the nearest tenth the answer is 5.3.

39.

$$
\begin{array}{r}
2.38 \\
24\overline{)57.14} \\
\underline{48} \\
91 \\
\underline{72} \\
194 \\
\underline{192} \\
2
\end{array}
$$

To the nearest tenth the answer is −2.4.

41.

$$
\begin{array}{r}
13.603 \\
9\overline{)122.430} \\
\underline{9} \\
32 \\
\underline{27} \\
54 \\
\underline{54} \\
030 \\
\underline{27} \\
3
\end{array}
$$

To the nearest hundredth the answer is 13.60.

43.

$$
\begin{array}{r}
0.791 \\
4\overline{)3.164} \\
\underline{28} \\
36 \\
\underline{36} \\
04 \\
\underline{4} \\
0
\end{array}
$$

To the nearest hundredth the answer is 0.79. -

45. $7.895 \div 100 = 0.07895$

47. $0.064 \div (-100) = -0.00064$

49.
$$1000\overline{)34.8000}^{\,0.0348}$$

51. $\dfrac{45.04}{10} = 4.504$

53.
$$\dfrac{-1.2 - 3.4}{3(1.6)} = \dfrac{-1.2 - 3.4}{4.8}$$
$$= \dfrac{-4.6}{4.8}$$
$$= -0.96$$

55.
$$\dfrac{40.7(-5.3)}{0.4 - 0.61} = \dfrac{-215.71}{-0.21}$$
$$= 1,027.19$$

57.
$$\dfrac{5(F - 32)}{9} = \dfrac{5(48.38 - 32)}{9}$$
$$= \dfrac{5(16.38)}{9}$$
$$= \dfrac{81.9}{9}$$
$$= 9.1$$

59.
$$\dfrac{6.7 - x^2 + 1.6}{x^3} = \dfrac{6.7 - (0.3)^2 + 1.6}{(0.3)^3}$$
$$= \dfrac{6.7 - 0.09 + 1.6}{0.027}$$
$$= \dfrac{8.21}{0.027}$$
$$= 304.07$$

APPLICATIONS

61.
$$\dfrac{0.68 + 0.36 + 0.44}{4} = \dfrac{1.48}{4}$$
$$= 0.37$$
The average depth is 0.37 mi.

63.
$$\dfrac{14}{.05} = \dfrac{1400}{5}$$
$$= 280$$
There will be 280 slices.

65.
$$\dfrac{27.5}{2.5} = \dfrac{275}{25}$$
$$= 11$$
It will take the hiker 11 hours so the arrival time will be 6 p.m.

67.
$$\dfrac{8.5}{0.015} = \dfrac{8500}{15}$$
$$\approx 567$$
There would be about 567 squeezes.

69. $451.20 \div 40 = 11.28$
$382.80 \div 40 = 9.57$
$549.60 \div 40 = 13.74$
A cook earns $11.28 per hour;
a server earns $9.57 per hour;
a manager earns $13.74 per hour.

71. $\dfrac{8.6 + 7.99 + (4.05)^2}{4.56} \approx 7.24$

73. $\left(\dfrac{45.9098}{-234.12}\right)^2 - 4 \approx -3.96$

REVIEW

79.

$$\frac{\frac{7}{8}}{\frac{3}{4}} = \frac{7}{8} \div \frac{3}{4}$$

$$= \frac{7}{8} \cdot \frac{4}{3}$$

$$= \frac{7 \cdot \overset{1}{4}}{2 \cdot 4 \cdot 3}$$

$$= \frac{7}{6}$$

81. Integers: ... ,–3, –2, –1, 0, 1, 2, 3, ...

83.

$$-\frac{3}{4}A = -9$$

$$-\frac{4}{3}\left(-\frac{3}{4}A\right) = -\frac{4}{3}(-9)$$

$$1A = -\frac{4}{3} \cdot -\frac{9}{1}$$

$$A = \frac{4 \cdot 3 \cdot \overset{1}{3}}{\underset{1}{3} \cdot 1}$$

$$A = 12$$

85. $5x - 6(x-1) - (-x) = 5x - 6x + 6 + x$
$$ = 6$$

STUDY SET Section 5.5

VOCABULARY

1. The decimal form of the fraction $\frac{1}{3}$ is a <u>repeating</u> decimal, which is written $0.\overline{3}$ or 0.3333. . . .

3. The set of decimals is called the set of <u>real</u> numbers.

CONCEPTS

5. $\frac{7}{8}$ can be interpreted as a fraction and as the division $7 \div 8$.

7. When rounding 0.272727... to the nearest hundredth, the result is smaller than the original number.

9.

11.

NOTATION

13. It is a repeating decimal.

15.
$$
\begin{array}{r}
0.5 \\
2\overline{)1.0} \\
\underline{10} \\
0
\end{array}
$$

$$\frac{1}{2} = 0.5$$

17.
$$
\begin{array}{r}
0.625 \\
8\overline{)5.000} \\
\underline{48} \\
20 \\
\underline{16} \\
40 \\
\underline{40} \\
0
\end{array}
$$

$$\frac{-5}{8} = -0.625$$

19.
$$
\begin{array}{r}
0.5625 \\
16\overline{)9.0000} \\
\underline{80} \\
100 \\
\underline{96} \\
40 \\
\underline{32} \\
80 \\
\underline{80} \\
0
\end{array}
$$

$$\frac{9}{16} = 0.5625$$

21.
$$
\begin{array}{r}
0.53125 \\
32\overline{)17.00000} \\
\underline{160} \\
100 \\
\underline{96} \\
40 \\
\underline{32} \\
80 \\
\underline{64} \\
160 \\
\underline{160} \\
0
\end{array}
$$

$$-\frac{17}{32} = -0.53125$$

23.

$$20\overline{)11.00} \atop 0.55$$

$$\begin{array}{r} 0.55 \\ 20\overline{)11.00} \\ \underline{100} \\ 100 \\ \underline{100} \\ 0 \end{array}$$

$$\frac{11}{20} = 0.55$$

25.

$$\begin{array}{r} 0.775 \\ 40\overline{)31.000} \\ \underline{280} \\ 300 \\ \underline{280} \\ 200 \\ \underline{200} \\ 0 \end{array}$$

$$\frac{31}{40} = 0.775$$

27.

$$\begin{array}{r} 0.015 \\ 200\overline{)3.000} \\ \underline{200} \\ 1000 \\ \underline{1000} \\ 0 \end{array}$$

$$-\frac{3}{200} = -0.015$$

29.

$$\begin{array}{r} 0.002 \\ 500\overline{)1.000} \\ \underline{1000} \\ 0 \end{array}$$

$$\frac{1}{500} = 0.002$$

31.

$$\begin{array}{r} 0.66 \\ 3\overline{)2.00} \\ \underline{18} \\ 20 \\ \underline{18} \\ 2 \end{array}$$

$$\frac{2}{3} = 0.\overline{6}$$

33.

$$\begin{array}{r} 0.454 \\ 11\overline{)5.000} \\ \underline{44} \\ 60 \\ \underline{55} \\ 50 \\ \underline{44} \\ 6 \end{array}$$

$$\frac{5}{11} = 0.\overline{45}$$

35.

$$\begin{array}{r} 0.5833 \\ 12\overline{)7.0000} \\ \underline{60} \\ 100 \\ \underline{96} \\ 40 \\ \underline{36} \\ 40 \\ \underline{36} \\ 4 \end{array}$$

$$-\frac{7}{12} = -0.58\overline{3}$$

37.

$$
\begin{array}{r}
0.033 \\
30\overline{)1.000} \\
\underline{90} \\
100 \\
\underline{90} \\
10
\end{array}
$$

$$\frac{1}{30} = 0.0\overline{3}$$

39.

$$
\begin{array}{r}
0.233 \\
30\overline{)7.000} \\
\underline{60} \\
100 \\
\underline{90} \\
100 \\
\underline{90} \\
10
\end{array}
$$

$$\frac{7}{30} \approx 0.23$$

41.

$$
\begin{array}{r}
0.377 \\
45\overline{)17.000} \\
\underline{135} \\
350 \\
\underline{315} \\
350 \\
\underline{315} \\
35
\end{array}
$$

$$\frac{17}{45} \approx 0.38$$

43.

$$
\begin{array}{r}
0.1515 \\
33\overline{)5.0000} \\
\underline{33} \\
170 \\
\underline{165} \\
50 \\
\underline{33} \\
170 \\
\underline{165} \\
5
\end{array}
$$

$$\frac{5}{33} \approx 0.152$$

45.

$$
\begin{array}{r}
0.3703 \\
27\overline{)10.0000} \\
\underline{81} \\
190 \\
\underline{189} \\
100 \\
\underline{81} \\
19
\end{array}
$$

$$\frac{10}{27} \approx 0.370$$

47.

$$
\begin{array}{r}
1.333 \\
3\overline{)4.000} \\
\underline{3} \\
10 \\
\underline{9} \\
10 \\
\underline{9} \\
10 \\
\underline{9} \\
1
\end{array}
$$

$$\frac{4}{3} \approx 1.33$$

49.

$$\begin{array}{r} 3.090 \\ 11\overline{)34.000} \\ \underline{33} \\ 100 \\ \underline{99} \\ 10 \end{array}$$

$$-\frac{34}{11} \approx -3.09$$

51.

$$\begin{array}{r} 0.75 \\ 4\overline{)3.00} \\ \underline{28} \\ 20 \\ \underline{20} \\ 0 \end{array}$$

$$3\frac{3}{4} = 3.75$$

53.

$$\begin{array}{r} 0.666 \\ 3\overline{)2.000} \\ \underline{18} \\ 20 \\ \underline{18} \\ 20 \\ \underline{18} \\ 2 \end{array}$$

$$-8\frac{2}{3} \approx -8.67$$

55.

$$\begin{array}{r} 0.6875 \\ 16\overline{)11.0000} \\ \underline{96} \\ 140 \\ \underline{128} \\ 120 \\ \underline{112} \\ 80 \\ \underline{80} \\ 0 \end{array}$$

$$12\frac{11}{16} = 12.6875$$

57.

$$\begin{array}{r} 0.733 \\ 15\overline{)11.000} \\ \underline{105} \\ 50 \\ \underline{45} \\ 50 \\ \underline{45} \\ 5 \end{array}$$

$$203\frac{11}{15} \approx 203.73$$

59. $\dfrac{7}{8} = 0.875$

$$\frac{7}{8} \leq 0.895$$

61. $\dfrac{11}{20} = 0.55$

$$-\frac{11}{20} \leq -0.\overline{4}$$

63. $-\dfrac{3}{4} \bullet 5.1 = -\dfrac{3}{4} \bullet \dfrac{51}{10}$

$$= -\frac{3 \bullet 51}{4 \bullet 10}$$

$$= -\frac{153}{40}$$

65. $\dfrac{1}{9} + 0.3 = \dfrac{1}{9} + \dfrac{3}{10}$

$$= \dfrac{1 \cdot 10}{9 \cdot 10} + \dfrac{3 \cdot 9}{10 \cdot 9}$$

$$= \dfrac{10}{90} + \dfrac{27}{90}$$

$$= \dfrac{37}{90}$$

67. $\dfrac{5}{11}(0.3) = \dfrac{5}{11} \cdot \dfrac{3}{10}$

$$= \dfrac{5 \cdot 3}{11 \cdot 10}$$

$$= \dfrac{\overset{1}{\cancel{5}} \cdot 3}{11 \cdot \underset{1}{\cancel{5}} \cdot 2}$$

$$= \dfrac{3}{22}$$

69. $\dfrac{1}{3}\left(-\dfrac{1}{15}\right)(0.5) = \dfrac{1}{3} \cdot -\dfrac{1}{15} \cdot \dfrac{5}{10}$

$$= -\dfrac{1 \cdot 1 \cdot 5}{3 \cdot 15 \cdot 10}$$

$$= -\dfrac{1 \cdot 1 \cdot \overset{1}{\cancel{5}}}{3 \cdot 5 \cdot 3 \cdot \underset{1}{\cancel{5}} \cdot 2}$$

$$= -\dfrac{1}{90}$$

71. $(3.5 + 6.7)\left(-\dfrac{1}{4}\right) = (3.5 + 6.7)(-0.25)$

$$= (10.2)(-0.25)$$

$$= -2.55$$

73. $\left(\dfrac{1}{5}\right)^2 (1.7) = (0.2)^2 (1.7)$

$$= (0.04)(1.7)$$

$$= 0.068$$

75. $7.5 - (0.78)\left(\dfrac{1}{2}\right) = 7.5 - (0.78)(0.5)$

$$= 7.5 - 0.39$$

$$= 7.11$$

77. $\dfrac{3}{8}(-3.2)+(4.5)\left(-\dfrac{1}{9}\right)=\left(\dfrac{3}{8}\right)\left(-\dfrac{3.2}{1}\right)+\left(\dfrac{4.5}{1}\right)\left(-\dfrac{1}{9}\right)$

$$=-\dfrac{9.6}{8}+\left(-\dfrac{4.5}{9}\right)$$
$$=-1.2+(-0.5)$$
$$=-1.7$$

79. $\dfrac{4}{3}pr^3=\left(\dfrac{4}{3}\right)(3.14)(3)^3$

$$=\left(\dfrac{4}{3}\right)\left(\dfrac{3.14}{1}\right)\left(\dfrac{27}{1}\right)$$
$$=\dfrac{12.56}{3}\left(\dfrac{27}{1}\right)$$
$$=\dfrac{339.12}{3}$$
$$=113.04$$

81. $\dfrac{23}{101}=0.\overline{2277}$

83. $\dfrac{1,736}{50}=34.72$

APPLICATIONS

85. $\dfrac{1}{16}=0.0625$; $\dfrac{6}{16}=0.375$;

$\dfrac{9}{16}=0.5625$; $\dfrac{15}{16}=0.9375$

87. $\dfrac{3}{40}=0.075$ so the $\dfrac{3}{40}$ in. line is thicker.

89. $23^2=23\dfrac{2}{5}=23.4$ sec

$23^4=23\dfrac{4}{5}=23.8$ sec

$24^1=24\dfrac{1}{5}=24.2$ sec

$32^3=32\dfrac{3}{5}=32.6$ sec

91. $A=6\left(\dfrac{1}{2}bh\right)$

$$=6\left[\dfrac{1}{2}(6)(5.2)\right]$$
$$=6\left(\dfrac{31.2}{2}\right)$$
$$=6(15.6)$$
$$=93.6$$

The area is 93.6 in.^{2}

REVIEW

95. $-2+(-3)+10+(-6)=-1$

97. $3T - 4T + 2(-4t) = 3T - 4T + (-8t)$
$$= -T - 8t$$

99. $4x^2 + 2x^2 = 6x^2$

STUDY SET Section 5.6

VOCABULARY

1. To <u>solve</u> an equation, we isolate the variable on one side of the equals sign.

3. In the term $5.65t$, 5.65 is called the numerical <u>coefficient</u>.

CONCEPTS

5. $2.1x - 6.3 = -2.73$

$2.1(1.7) - 6.3 \overset{?}{=} -2.73$

$3.57 - 6.3 \overset{?}{=} -2.73$

$-2.73 = -2.73$

7. Simplify applies to $7.8x + 9.1 + 3.2x$

9. a. 25 cents = $0.25

   b. 1 penny = $0.01

   c. 250 cents = $2.50

11. The distributive property is being shown.

NOTATION

13. $0.6s - 2.3 = -1.82$

$0.6s - 2.3 + \underline{2.3} = -1.82 + \underline{2.3}$

$\underline{0.6s} = 0.48$

$\dfrac{0.6s}{0.6} = \dfrac{0.48}{0.6}$

$s = 0.8$

PRACTICE

15. $8.7x + 1.4x = 10.1x$

17. $0.05h - 0.03h = 0.02h$

19. $3.1r - 5.5r - 1.3r = -3.7r$

21. $3.2 - 8.78x + 9.1 = -8.78x + 12.3$

23. $5.6x - 8.3 - 6.1x + 12.2 = -0.5x + 3.9$

25. $0.05(100 - x) + 0.04x = 5 - 0.05x + 0.04x$
$$= -0.01x + 5$$

27.
$$x + 8.1 = 9.8$$
$$x + 8.1 - 8.1 = 9.8 - 8.1$$
$$x = 1.7$$

29.
$$7.08 = t - 0.03$$
$$7.08 + 0.03 = t - 0.03 + 0.03$$
$$7.11 = t$$

31.
$$-5.6 + h = -17.1$$
$$-5.6 + h + 5.6 = -17.1 + 5.6$$
$$h = -11.5$$

33.
$$7.75 = t - (-7.85)$$
$$7.75 = t + 7.85$$
$$7.75 - 7.85 = t + 7.85 - 7.85$$
$$-0.1 = t$$

35. $2x = -8.72$
$$\frac{2x}{2} = \frac{-8.72}{2}$$
$$x = -4.36$$

37. $-3.51 = -2.7x$
$$\frac{-3.51}{-2.7} = \frac{-2.7x}{-2.7}$$
$$1.3 = x$$

39.
$$\frac{x}{2.04} = -4$$
$$2.04\left(\frac{x}{2.04}\right) = 2.04(-4)$$
$$x = -8.16$$

41.
$$\frac{-x}{5.1} = -4.4$$
$$-5.1\left(\frac{-x}{5.1}\right) = -5.1(-4.4)$$
$$x = 22.44$$

43. $\frac{1}{3}x = -7.06$
$$3\left(\frac{1}{3}x\right) = 3(-7.06)$$
$$x = -21.18$$

45.
$$\frac{x}{100} = 0.004$$
$$100\left(\frac{x}{100}\right) = 100(0.004)$$
$$x = 0.4$$

47.
$$2x + 7.8 = 3.4$$
$$2x + 7.8 - 7.8 = 3.4 - 7.8$$
$$2x = -4.4$$
$$\frac{2x}{2} = \frac{-4.4}{2}$$
$$x = -2.2$$

49.
$$-0.8 = 5y + 9.2$$
$$-0.8 - 9.2 = 5y + 9.2 - 9.2$$
$$-10 = 5y$$
$$\frac{-10}{5} = \frac{5y}{5}$$
$$-2 = y$$

51.
$$0.3x - 2.1 = 7.2$$
$$0.3x - 2.1 + 2.1 = 7.2 + 2.1$$
$$0.3x = 9.3$$
$$\frac{0.3x}{0.3} = \frac{9.3}{0.3}$$
$$x = 31$$

53.
$$-1.5b + 2.7 = 1.2$$
$$-1.5b + 2.7 - 2.7 = 1.2 - 2.7$$
$$-1.5b = -1.5$$
$$\frac{-1.5b}{-1.5} = \frac{-1.5}{-1.5}$$
$$b = 1$$

55.
$$0.4a - 6 + 0.5a = -5.730$$
$$0.9a - 6 = -5.73$$
$$0.9a - 6 + 6 = -5.73 + 6$$
$$0.9a = 0.27$$
$$\frac{0.9a}{0.9} = \frac{0.27}{0.9}$$
$$a = 0.3$$

57.
$$2(t - 4.3) + 1.2 = -6.2$$
$$2t - 8.6 + 1.2 = -6.2$$
$$2t - 7.4 = -6.2$$
$$2t - 7.4 + 7.4 = -6.2 + 7.4$$
$$2t = 1.2$$
$$\frac{2t}{2} = \frac{1.2}{2}$$
$$t = 0.6$$

59.
$$1.2x - 1.3 = 2.4x + 0.02$$
$$1.2x - 1.3 - 1.2x = 2.4x + 0.02 - 1.2x$$
$$-1.3 = 1.2x + 0.02$$
$$-1.3 - 0.02 = 1.2x + 0.02 - 0.02$$
$$-1.32 = 1.2x$$
$$\frac{-1.32}{1.2} = \frac{1.2x}{1.2}$$
$$-1.1 = x$$

61.
$$53.7t - 10.1 = 46.3t + 4.7$$
$$53.7t - 10.1 - 46.3t = 46.3t + 4.7 - 46.3t$$
$$7.4t - 10.1 = 4.7$$
$$7.4t - 10.1 + 10.1 = 4.7 + 10.1$$
$$7.4t = 14.8$$
$$\frac{7.4t}{7.4} = \frac{14.8}{7.4}$$
$$t = 2$$

63.
$$2.1x - 4.6 = 7.3x - 11.36$$
$$2.1x - 4.6 - 2.1x = 7.3x - 11.36 - 2.1x$$
$$-4.6 = 5.2x - 11.36$$
$$-4.6 + 11.36 = 5.2x - 11.36 + 11.36$$
$$6.76 = 5.2x$$
$$\frac{6.76}{5.2} = \frac{5.2x}{5.2}$$
$$1.3 = x$$

65.
$$0.06x + 0.09(100 - x) = 8.85$$
$$0.06x + 9 - 0.09x = 8.85$$
$$-0.03x + 9 = 8.85$$
$$-0.03x + 9 - 9 = 8.85 - 9$$
$$-0.03x = -0.15$$
$$\frac{-0.03x}{-0.03} = \frac{-0.15}{-0.03}$$
$$x = 5$$

APPLICATIONS

67. **A:** <u>The number of signatures she needs</u>
  <u>to collect</u>

  F: Let x = <u>the number of signatures</u>
  <u>she needs to collect</u>
  We need to work in terms of the same
  units, so we write 30 cents as <u>$0.30.</u>
  <u>$0.30x</u> = total amount made from
  collecting signatures

$$\boxed{\textbf{15}} + \underline{0.30} \cdot \boxed{\begin{array}{c}\textbf{the number} \\ \textbf{of signatures}\end{array}} \text{ is 60.}$$

$$15 + \underline{0.30x} = 60$$

S: $15 + \underline{0.30x} = \underline{60}$

$$0.30x = 45$$

$$x = 150$$

S: She needs to collect 150 signatures to make $60.

C: If she collects $\underline{150}$ signatures, she will make $\underline{0.30} \cdot \underline{150} = \underline{45}$ dollars from signatures. If we add this to $15, we get $60. The answer checks.

69. Let x = the amount the county should ask for

$$6.8 + 12.5 + x = 27.9$$

$$19.3 + x = 27.9$$

$$19.3 + x - 19.3 = 27.9 - 19.3$$

$$x = 8.6$$

The county should ask for $8.6 million.

71. Let x = her GPA at the beginning of the fall semester

$$x - 0.18 = 3.09$$

$$x - 0.18 + 0.18 = 3.09 + 0.18$$

$$x = 3.27$$

Her GPA at the beginning of the fall semester was 3.27.

73. Let x = the amount of each monthly payment

$$3x = 113.25$$

$$\frac{3x}{3} = \frac{113.25}{3}$$

$$x = 37.75$$

Each monthly payment is $37.75.

75. Let x = the maximum number of words that can be printed on the award

$$20 + 0.15x = 50$$

$$20 + 0.15x - 20 = 50 - 20$$

$$0.15x = 30$$

$$\frac{0.15x}{0.15} = \frac{30}{0.15}$$

$$x = 200$$

The maximum number of words is 200.

REVIEW

79. $-\dfrac{2}{3} + \dfrac{3}{4} = -\dfrac{2 \cdot 4}{3 \cdot 4} + \dfrac{3 \cdot 3}{4 \cdot 3}$

$$= -\frac{8}{12} + \frac{9}{12}$$

$$= \frac{1}{12}$$

81.
$$x^3 - y^3 = \left(-\frac{1}{2}\right)^3 - (-1)^3$$
$$= -\frac{1}{8} - (-1)$$
$$= -\frac{1}{8} + 1$$
$$= -\frac{1}{8} + \frac{8}{8}$$
$$= \frac{7}{8}$$

83.
$$\frac{-3-3}{-3+4} = \frac{-6}{1}$$
$$= -6$$

STUDY SET Section 5.7

VOCABULARY

1. When we find what number is squared to obtain a given number, we are finding the <u>square root</u> of the given number.

3. The symbol $\sqrt{}$ is called a <u>radical sign</u>. It indicates that we are to find a <u>positive</u> square root.

5. In $\sqrt{26}$, 26 is called the <u>radicand</u>.

19.

21. a. $\sqrt{19}$ would be between 4 and 5.

　b. $\sqrt{87}$ would be between 9 and 10.

NOTATION

23. $-\sqrt{49} + \sqrt{64} = \underline{-7} + \underline{8}$
$$= 1$$

CONCEPTS

7. The square of 5 is _25_, because $(5)^2 =$ _25._

9. The two square roots of 49 are 7 and –7, because $\underline{7}^2 = 49$ and $\underline{(-7)}^2 = 49$.

11. Since $\left(\frac{3}{4}\right)^2 = \frac{9}{16}, \sqrt{\frac{9}{16}} = \frac{3}{4}$

13. Smallest to largest:
$$\sqrt{6}, \sqrt{11}, \sqrt{23}, \sqrt{27}$$

15. a. $\sqrt{1} = 1$

　b. $\sqrt{0} = 0$

17. a. $\sqrt{6} \approx 2.4$

　b. $(2.4)^2 = 5.76$

　c. $6 - 5.76 = 0.24$

PRACTICE

25. $\sqrt{16} = 4$

27. $-\sqrt{121} = -11$

29. $-\sqrt{0.49} = -0.7$

31. $\sqrt{0.25} = 0.5$

33. $\sqrt{0.09} = 0.3$

35. $-\sqrt{\dfrac{1}{81}} = -\dfrac{1}{9}$

37. $-\sqrt{\dfrac{16}{9}} = -\dfrac{4}{3}$

39. $\sqrt{\dfrac{4}{25}} = \dfrac{2}{5}$

41. $5\sqrt{36} + 1 = 5(6) + 1$
$\qquad = 30 + 1$
$\qquad = 31$

43. $-4\sqrt{36} + 2\sqrt{4} = -4(6) + 2(2)$
$\qquad\qquad\qquad = -24 + 4$
$\qquad\qquad\qquad = -20$

45. $\sqrt{\dfrac{1}{16}} - \sqrt{\dfrac{9}{25}} = \dfrac{1}{4} - \dfrac{3}{5}$
$\qquad\qquad\quad = \dfrac{1 \cdot 5}{4 \cdot 5} - \dfrac{3 \cdot 4}{5 \cdot 4}$
$\qquad\qquad\quad = \dfrac{5}{20} - \dfrac{12}{20}$
$\qquad\qquad\quad = -\dfrac{7}{20}$

47. $5(\sqrt{49})(-2) = 5(7)(-2)$
$\qquad\qquad\quad = 35(-2)$
$\qquad\qquad\quad = -70$

49. $\sqrt{0.04} + 2.36 = 0.2 + 2.36$
$\qquad\qquad\quad = 2.56$

51. $-3\sqrt{1.44} = -3(1.2)$
$\qquad\qquad = -3.6$

53.

Number	Square root
1	**1**
2	**1.414**
3	**1.732**
4	**2**
5	**2.236**
6	**2.449**
7	**2.646**
8	**2.828**
9	**3**
10	**3.162**

55. $\sqrt{1,369} = 37$

57. $\sqrt{3,721} = 61$

59. $\sqrt{15} \approx 3.87$

61. $\sqrt{66} \approx 8.12$

63. $\sqrt{24.05} \approx 4.904$

65. $-\sqrt{11.1} \approx -3.332$

APPLICATIONS

67. a. $\sqrt{25} = 5$ The length is 5 ft.

67. b. $\sqrt{100} = 10$ The length is 10 ft.

69. $\sqrt{16,200} \approx 127.3$
The distance from home plate to second base is about 127.3 ft.

71. $\sqrt{1,681} = 41$ A 41-inch screen is shown.

73. $\sqrt{24,000,201} = 4,899$

75. $-\sqrt{0.00111} = -0.0333$

REVIEW

81. To isolate the variable, subtraction and multiplication must be undone.

83. $5(-2)^2 - \dfrac{16}{4} = 5(4) - \dfrac{16}{4}$
$= 20 - 4$
$= 16$

85. Whole numbers: 0, 1, 2, 3, 4, 5, 6, . . .

87. $8 + \dfrac{a}{5} = 14$

$8 + \dfrac{a}{5} - 8 = 14 - 8$

$\dfrac{a}{5} = 6$

$5\left(\dfrac{a}{5}\right) = 5(6)$

$a = 30$

CHAPTER 5 REVIEW

1. The shaded amount is 0.67 or $\dfrac{67}{100}$.

3. $16.4523 = 10 + 6 + \dfrac{4}{10} + \dfrac{5}{100} + \dfrac{2}{1,000} + \dfrac{3}{10,000}$

5.

7. Washington, Diaz, Chou, Singh, Gerbac

9. a. $4.5 \leq 4.6$

  b. $-2.35 \geq -2.53$

  c. $10.90 = 10.9$

  d. $0.027894 \leq 0.034$

11. a. $19.5 + 34.4 + 12.8 = 66.7$

  b. $3.4 + 6.78 + 35 + 0.008 = 45.188$

13. a. $-16.1 + 8.4 = -7.7$

 b. $-4.8 - (-7.9) = 3.1$

 c. $-3.55 + (-1.25) = -4.8$

 d. $-15.1 - 13.99 = -29.09$

15. $52.20 - 3.99 = 48.21$
 The sale price is \$48.21.

17. a. $(-0.6)(0.4) = -0.24$

 b. $2.3 \cdot 0.9 = 2.07$

 c. $5.5(-3.1) = -17.05$

 d. $32.45(6.1) = 197.945$

17. e. $(-0.003)(-0.02) = 0.00006$

 f. $7 \cdot 0.6 = 4.2$

19. a. $(0.2)^2 = 0.04$

 b. $(-0.15)^2 = 0.0225$

 c. $(3.3)^2 = 10.89$

 d. $(0.1)^3 = 0.001$

21. $\begin{aligned} 2pr^2 - h &= 2(3.14)(4)^2 - 8.1 \\ &= 2(3.14)(16) - 8.1 \\ &= 100.48 - 8.1 \\ &= 92.38 \end{aligned}$

23. $\begin{aligned} 0.03 + 3(0.015) - 0.005 &= 0.03 + 0.045 - 0.005 \\ &= 0.075 - 0.005 \\ &= 0.07 \end{aligned}$
 The thickness is 0.07 in.

25. a.

$$\begin{array}{r} 2.9 \\ 43\overline{)124.7} \\ \underline{86} \\ 387 \\ \underline{387} \\ 0 \end{array}$$

$12.47 \div (-4.3) = -2.9$

25. b.

$$14\overline{)0.742}$$ with quotient 0.053

$$
\begin{array}{r}
0.053 \\
14\overline{)0.742} \\
\underline{70} \\
42 \\
\underline{42} \\
0
\end{array}
$$

$$\frac{0.0742}{1.4} = 0.053$$

c.

$$
\begin{array}{r}
63 \\
25\overline{)1575} \\
\underline{150} \\
75 \\
\underline{75} \\
0
\end{array}
$$

$$\frac{15.75}{0.25} = 63$$

d.

$$
\begin{array}{r}
0.81 \\
46\overline{)37.26} \\
\underline{368} \\
46 \\
\underline{46} \\
0
\end{array}
$$

$$\frac{-0.03726}{-0.046} = 0.81$$

27.

$$
\begin{aligned}
\frac{5(F-32)}{9} &= \frac{5(68.4-32)}{9} \\
&= \frac{5(36.4)}{9} \\
&= \frac{182}{9} \\
&= 20.\overline{2} \\
&\approx 20.22
\end{aligned}
$$

29. a. $89.76 \div 100 = 0.8976$

b. $\dfrac{0.0112}{-10} = -0.00112$

31.

$$
\begin{array}{r}
14.09 \\
11\overline{)155.00} \\
\underline{11} \\
45 \\
\underline{44} \\
100 \\
\underline{99} \\
1
\end{array}
$$

There are 14 servings

33. a.

$$
\begin{array}{r}
0.875 \\
8\overline{)7.000} \\
\underline{64} \\
60 \\
\underline{56} \\
40 \\
\underline{40} \\
0
\end{array}
$$

$$\frac{7}{8} = 0.875$$

b.

$$
\begin{array}{r}
0.4 \\
5\overline{)2.0} \\
\underline{20} \\
0
\end{array}
$$

$$-\frac{2}{5} = -0.4$$

33. c.

$$
\begin{array}{r}
0.5625 \\
16\overline{)9.0000} \\
\underline{80} \\
100 \\
\underline{96} \\
40 \\
\underline{32} \\
80 \\
\underline{80} \\
0
\end{array}
$$

$$\frac{9}{16} = 0.5625$$

d.

$$
\begin{array}{r}
0.06 \\
50\overline{)3.00} \\
\underline{300} \\
0
\end{array}
$$

$$\frac{3}{50} = 0.06$$

35. a.

$$
\begin{array}{r}
0.575 \\
33\overline{)19.000} \\
\underline{165} \\
250 \\
\underline{231} \\
190 \\
\underline{165} \\
25
\end{array}
$$

$$\frac{19}{33} \approx 0.58$$

35. b.

$$
\begin{array}{r}
1.033 \\
30\overline{)31.000} \\
\underline{30} \\
100 \\
\underline{90} \\
100 \\
\underline{90} \\
10
\end{array}
$$

$$\frac{31}{30} \approx 1.03$$

37.

39. $\dfrac{4}{3}(3.14)(2)^3 = \dfrac{4}{3}(3.14)(8)$

$$= \dfrac{4}{3}(25.12)$$

$$= \dfrac{4}{3}\left(\dfrac{25.12}{1}\right)$$

$$= \dfrac{100.48}{3}$$

$$\approx 33.49$$

41. a.
$$y + 12.4 = -6.01$$
$$y + 12.4 - 12.4 = -6.01 - 12.4$$
$$y = -18.41$$

b.
$$0.23 + x = 5$$
$$0.23 + x - 0.23 = 5 - 0.23$$
$$x = 4.77$$

c.
$$\dfrac{x}{1.78} = -3$$
$$1.78\left(\dfrac{x}{1.78}\right) = 1.78(-3)$$
$$x = -5.34$$

d. $-16.1b = -27.37$
$$\dfrac{-16.1b}{-16.1} = \dfrac{-27.37}{-16.1}$$
$$b = 1.7$$

43. a. $5.7a - 12.4 - 2.9a = 2.8a - 12.4$

b. $2(0.3t - 0.4) + 3(0.8t - 0.2) = 0.6t - 0.8 + 2.4t - 0.6$
$$= 3t - 1.4$$

45. Let x = the number of games that can be bowled

$$1.45 + 0.95x = 10$$
$$1.45 + 0.95x - 1.45 = 10 - 1.45$$
$$0.95x = 8.55$$
$$\frac{0.95x}{0.95} = \frac{8.55}{0.95}$$
$$x = 9$$

Nine games can be bowled for $10.

47. a. $\sqrt{49} = 7$

b. $-\sqrt{16} = -4$

c. $\sqrt{100} = 10$

d. $\sqrt{0.09} = 0.3$

e. $\sqrt{\dfrac{64}{25}} = \dfrac{8}{5}$

f. $\sqrt{0.81} = 0.9$

g. $-\sqrt{\dfrac{1}{36}} = -\dfrac{1}{6}$

h. $\sqrt{0} = 0$

49. $\sqrt{11} \approx 3.31662479$
$$\approx 3.3$$
$$(3.3)^2 = 10.89$$
$$11 - 10.89 = 0.11$$
The difference is 0.11.

51. a. $-3\sqrt{100} = -3(10)$
$$= -30$$

b. $5\sqrt{0.25} = 5(0.5$
$$= 2.5$$

c. $-3\sqrt{49} - \sqrt{36} = -3(7) - 6$
$$= -21 - 6$$
$$= -27$$

d. $\sqrt{\dfrac{9}{100}} + \sqrt{1.44} = \dfrac{3}{10} + 1.2$
$$= 0.3 + 1.2$$
$$= 1.5$$

KEY CONCEPT Simplify and Solve

1. a. $-2x + 3 + 7x - 11 = -2x + 7x + 3 - 11$
$$= 5x - 8$$

1. b.
$$-2x + 3 + 7x - 11 = 7$$
$$5x - 8 = 7$$
$$5x - 8 + 8 = 7 + 8$$
$$5x = 15$$
$$\frac{5x}{5} = \frac{15}{5}$$
$$x = 3$$

3. a.
$$3(0.2y - 1.6) + 0.6y = 0.6y - 4.8 + 0.6y$$
$$= 1.2y - 4.8$$

b.
$$3(0.2y - 1.6) + 0.6y = -6.6$$
$$0.6y - 4.8 + 0.6y = -6.6$$
$$1.2y - 4.8 = -6.6$$
$$1.2y - 4.8 + 4.8 = -6.6 + 4.8$$
$$1.2y = -1.8$$
$$\frac{1.2y}{1.2} = \frac{-1.8}{1.2}$$
$$y = -1.5$$

KEY CONCEPT The Real Numbers

1. Natural numbers: 1, 2, 3, 4, 5, . .

3. Integers: . . ., –3, –2, –1, 0, 1, 2, 3, . . .

5. Irrational numbers: nonterminating, nonrepeating decimals; a number that can't be written as a fraction

7. False; some can be written as a repeating decimal.

9. False; all integers are rational numbers.

11. True

13. False; the set of whole numbers is a subset of the rational numbers.

15. True

CHAPTER 5 TEST

1. The shaded amount is $\frac{79}{100}$ or 0.79.

3. $0.271 = \frac{271}{1,000}$

5. $(30.25 + 62.25) + (40.50 + 75.75) = 92.50 + 116.25$
$$= 208.75$$
The total income for the two days is $208.75.

7. $-0.83 + (-0.19) = -1.02$
The ground dropped 1.02 in.

9. $A = lw$
$$= 3.6(1.8)$$
$$= 6.48$$
The area is 6.48 mi^2.

11. $4.1 - (3.2)(0.4)^2 = 4.1 - (3.2)(0.16)$
$$= 4.1 - 0.512$$
$$= 3.588$$

13.
$$
\begin{array}{r}
2.291 \\
53\overline{)121.460} \\
\end{array}
$$

$$
\begin{array}{r}
2.291 \\
53\,\overline{)121.460} \\
\underline{106} \\
154 \\
\underline{106} \\
486 \\
\underline{477} \\
90 \\
\underline{53} \\
37 \\
\end{array}
$$

$$\frac{12.146}{-5.3} = -2.29$$

15.

17. $6.18s + 8.9 - 1.22s - 6.6 = 4.96s + 2.3$

19. a. $-2.4t = 16.8$
$$\frac{-2.4t}{-2.4} = \frac{16.8}{-2.4}$$
$$t = -7$$

b. $-0.008 + x = 6$
$$-0.008 + x + 0.008 = 6 + 0.008$$
$$x = 6.008$$

21. Let C = the weight of compound C
$$1.86 + 2.09 + C = 4.37$$
$$3.95 + C = 4.37$$
$$3.95 + C - 3.95 = 4.37 - 3.95$$
$$C = 0.42$$
Compound C weighed 0.42 g.

23.

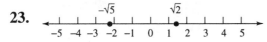

25. a. $-6.78 \underline{>} -6.79$

b. $\dfrac{3}{8} \underline{>} 0.3$

c. $\sqrt{\dfrac{16}{81}} \underline{>} \dfrac{16}{81}$

d. $0.\overline{45} \underline{>} 0.45$

STUDY SET Section 6.1

VOCABULARY

1. The pair of numbers (2, 5) is called an <u>ordered pair</u>.

3. The rectangular coordinate system is sometimes called the <u>Cartesian coordinate system</u>.

5. The coordinate axes that divide a rectangular coordinate system into quadrants are called the <u>x-axis</u> and the <u>y-axis</u>.

7. In the ordered pair (–2, 4), <u>–2</u> is the x-coordinate.

CONCEPTS

9. To plot the point with coordinates (3, –4), we start at the <u>origin</u> and move 3 units to the <u>right</u> and then move 4 units <u>down</u>.

11. a. On the x-axis, the grid lines are one unit apart.

    b. On the y-axis, the grid lines are 5 units apart.

NOTATION

13.
$$4x - 3y = 12$$
$$4(\underline{2}) - 3y = 12$$
$$\underline{8} - 3y = 12$$
$$-3y = \underline{4}$$
$$y = -\frac{4}{3}$$

PRACTICE

15. a.
$$3x + 4y = 12$$
$$3(0) + 4y = 12$$
$$4y = 12$$
$$y = 3$$
If $x = 0$, then $y = \underline{3}$.

    b.
$$3x + 4y = 12$$
$$3x + 4(0) = 12$$
$$3x = 12$$
$$x = 4$$
If $y = 0$, then $x = \underline{4}$.

    c.
$$3x + 4y = 12$$
$$3(2) + 4y = 12$$
$$6 + 4y = 12$$
$$4y = 6$$
$$y = \frac{3}{2}$$
If $x = 2$, then $y = \frac{3}{2}$.

17. a.
$$2x + y = 8$$
$$2(0) + y = 8$$
$$y = 8$$
(0, 8)

    b.
$$2x + y = 8$$
$$2x + 0 = 8$$
$$2x = 8$$
$$x = 4$$
(4, 0)

    c.
$$2x + y = 8$$
$$2x + 2 = 8$$
$$2x = 6$$
$$x = 3$$
(3, 2)

19. If we substitute 0 for x, we get $y = -5$. Enter -5 in the first row of the table. If we substitute 0 for y, we get $x = 4$. Enter 4 in the second row of the table. If we substitute 5 for y, we get $x = 8$. Enter 8 in the third row of the table.

x	y	(x, y)
0	-5	$(0, -5)$
4	0	$(4, 0)$
8	5	$(8, 5)$

21.

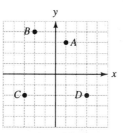

23.

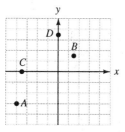

25. $A(2, 4)$, $B(-3, 3)$, $C(-2, -3)$, $D(4, -3)$

27. $A(-3, -4)$, $B\left(\dfrac{5}{2}, \dfrac{7}{2}\right)$, $C\left(-\dfrac{5}{2}, 0\right)$, $D\left(\dfrac{5}{2}, 0\right)$

APPLICATIONS

31. The damage extended 5 miles from the epicenter.

33.

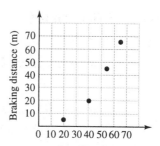

Braking distances are longer at greater speeds.

35.

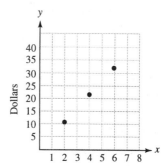

The student will earn approximately $38.50 in 7 hours.

37.

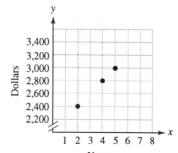

The account will have a value of approximately $3,200 in 6 years.

REVIEW

43. $(-8 - 5) - 3 = -13 - 3$
$= -16$

45. $(-4)^2 - 3^2 = 16 - 9$
$= 7$

47. $\dfrac{x}{3} + 3 = 10$

$\quad\quad \dfrac{x}{3} = 7$

$\quad\quad x = 21$

Check: $\dfrac{x}{3} + 3 = 10$

$\quad\quad \dfrac{21}{3} + 3 \overset{?}{=} 10$

$\quad\quad 7 + 3 \overset{?}{=} 10$

$\quad\quad 10 = 10$

49. $5 - (7 - x) = -5$

$\quad\quad 5 - 7 + x = -5$

$\quad\quad -2 + x = -5$

$\quad\quad x = -3$

Check: $5 - (7 - x) = -5$

$\quad\quad 5 - [7 - (-3)] \overset{?}{=} -5$

$\quad\quad 5 - 10 \overset{?}{=} -5$

$\quad\quad -5 = -5$

51. $(x^3)^4 = x^{3 \cdot 4}$

$\quad\quad = x^{12}$

53. $x^3 x^4 = x^{3+4}$

$\quad\quad = x^7$

55. $(4^2)^4 = 4^{2 \cdot 4}$

$\quad\quad = 4^8$

$\quad\quad = 65{,}536$

CONCEPTS

7. The graph of the equation $y = 3$ is a <u>horizontal</u> line.

9. a. The line crosses the y-axis at $y = 1$, so the y-intercept is $(0, 1)$.

b. The line crosses the x-axis at $x = -2$, so the x-intercept is $(-2, 0)$.

c. Yes

NOTATION

11. $\quad 2x - 4y = 8$

$\quad\quad 2(\underline{3}) - 4y = 8$

$\quad\quad \underline{6} - 4y = 8$

$\quad\quad -4y = \underline{2}$

$\quad\quad y = -\dfrac{1}{2}$

STUDY SET Section 6.2

VOCABULARY

1. The graph of a linear equation is a <u>line</u>.

3. The point where the graph of a linear equation crosses the x-axis is called the <u>x-intercept</u>.

5. In the equation $y = 7x + 2$, x is called the <u>independent</u> variable.

PRACTICE

13. $x = 5$:
$$2x - 5y = 10$$
$$2(5) - 5y = 10$$
$$10 - 5y = 10$$
$$-5y = 0$$
$$y = 0$$

$x = -5$:
$$2x - 5y = 10$$
$$2(-5) - 5y = 10$$
$$-10 - 5y = 10$$
$$-5y = 20$$
$$y = -4$$

$x = 10$:
$$2x - 5y = 10$$
$$2(10) - 5y = 10$$
$$20 - 5y = 10$$
$$-5y = -10$$
$$y = 2$$

x	y	(x, y)
5	0	$(5, 0)$
-5	-4	$(-5, -4)$
10	2	$(10, 2)$

15. $x = 3$:
$$y = 2x - 3$$
$$y = 2(3) - 3$$
$$y = 6 - 3$$
$$y = 3$$

$x = -4$:
$$y = 2x - 3$$
$$y = 2(-4) - 3$$
$$y = -8 - 3$$
$$y = -11$$

$x = 6$:
$$y = 2x - 3$$
$$y = 2(6) - 3$$
$$y = 12 - 3$$
$$y = 9$$

x	y	(x, y)
3	3	$(3, 3)$
-4	-11	$(-4, -11)$
6	9	$(6, 9)$

17. $x = 0$ at the y-intercept:
$$x + y = 5$$
$$0 + y = 5$$
$$y = 5$$
The y-intercept is $(0, 5)$.

$y = 0$ at the x-intercept:
$$x + y = 5$$
$$x + 0 = 5$$
$$x = 5$$
The x-intercept is $(5, 0)$.

19. $x = 0$ at the y-intercept:
$$4x + 5y = 20$$
$$4(0) + 5y = 20$$
$$5y = 20$$
$$y = 4$$
The y-intercept is $(0, 4)$.

$y = 0$ at the x-intercept:
$$4x + 5y = 20$$
$$4x + 5(0) = 20$$
$$4x = 20$$
$$x = 5$$
The x-intercept is $(5, 0)$.

21.

x	y
0	5
5	0
2	3

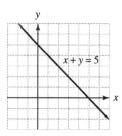

23.

x	y
0	4
5	0
1	$\frac{16}{5}$

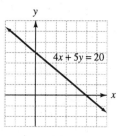

25.

x	y
0	2
−4	0
4	4

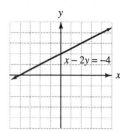

27.

x	y
0	−5
$\frac{5}{2}$	0
1	−3

Tables will vary.

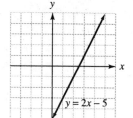

29.

x	y
0	2
$\frac{4}{3}$	0
2	−1

Tables will vary.

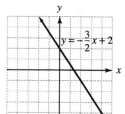

31.

x	y
−1	5
0	5
1	5

Tables will vary.

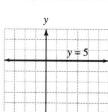

33.

x	y
4	−3
4	0
4	3

Tables will vary.

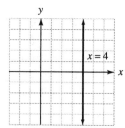

35.

x	y
−2	−4
0	0
2	4

Tables will vary.

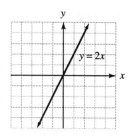

37.

x	y
−3	−1
0	0
3	1

Tables will vary.

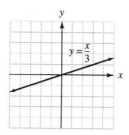

APPLICATIONS

39.

t	d
1	2
2	4
3	6
4	8
5	10

Tables will vary.

41. Yes, the lines both go through (2, 2), so a collision is possible at that point.

REVIEW

47. $180 = 2^2 \cdot 3^2 \cdot 5$

49. $3(x - 4) = 3 \cdot x - 3 \cdot 4$
$= 3x - 12$

51. $3a - 2b = 3(-2) - 2(3)$
$= -6 - 6$
$= -12$

53. $-3a + 4a^2 = -3(-2) + 4(-2)^2$
$= 6 + 4(4)$
$= 6 + 16$
$= 22$

55. $C = \dfrac{5(F - 32)}{9}$

$C = \dfrac{5(77 - 32)}{9}$

$= \dfrac{5(45)}{9}$

$= \dfrac{225}{9}$

$= 25$

77° F is 25° C.

STUDY SET Section 6.3

VOCABULARY

1. A polynomial with one algebraic term is called a <u>monomial</u>.

3. A polynomial with two algebraic terms is called a <u>binomial</u>.

CONCEPTS

5. $3x^2 - 4$ has two terms, so it is a binomial.

7. $17e^4$ has one term, so it is a monomial.

9. $25u^2$ has one term, so it is a monomial.

11. $q^5 + q^2 + 1$ has three terms, so it is a trinomial.

13. The degree of $5x^3$ is 3.

15. The degree of $2x^2$ is 2, the degree of $-3x$ is 1, and the degree of 2 is 0, so the degree of $2x^2 - 3x + 2$ is 2.

17. The degree of $2m$ is 1.

19. The degree of $25w^6$ is 6 and the degree of $5w^7$ is 7, so the degree of $25w^6 + 5w^7$ is 7.

NOTATION

21.
$$\begin{aligned}
3a^2 + 2a - 7 &= 3(\underline{2})^2 + 2(\underline{2}) - 7 \\
&= 3(\underline{4}) + \underline{4} - 7 \\
&= 12 + 4 - 7 \\
&= \underline{16} - 7 \\
&= 9
\end{aligned}$$

PRACTICE

23.
$$\begin{aligned}
3x + 4 &= 3(3) + 4 \\
&= 9 + 4 \\
&= 13
\end{aligned}$$

25.
$$\begin{aligned}
2x^2 + 4 &= 2(-1)^2 + 4 \\
&= 2(1) + 4 \\
&= 2 + 4 \\
&= 6
\end{aligned}$$

27.
$$\begin{aligned}
0.5t^3 - 1 &= 0.5(4)^3 - 1 \\
&= 0.5(64) - 1 \\
&= 32 - 1 \\
&= 31
\end{aligned}$$

29.
$$\begin{aligned}
\frac{2}{3}b^2 - b + 1 &= \frac{2}{3}(3)^2 - (3) + 1 \\
&= \frac{2}{3}(9) - 3 + 1 \\
&= 6 - 3 + 1 \\
&= 3 + 1 \\
&= 4
\end{aligned}$$

31.
$$\begin{aligned}
-2s^2 - 2s + 1 &= -2(-1)^2 - 2(-1) + 1 \\
&= -2(1) + 2 + 1 \\
&= -2 + 2 + 1 \\
&= 0 + 1 \\
&= 1
\end{aligned}$$

33.

x	y	(x, y)
-2	4	(-2, 4)
-1	1	(-1, 1)
0	0	(0, 0)
1	1	(1, 1)
2	4	(2, 4)

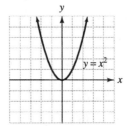

35.

x	y	(x, y)
-2	2	(-2, 2)
0	0	(0, 0)
2	2	(2, 2)

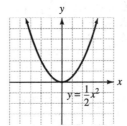

37.

x	y	(x, y)
-2	-3	(-2, -3)
-1	0	(-1, 0)
0	1	(0, 1)
1	0	(1, 0)
2	-3	(2, -3)

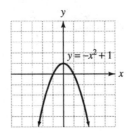

39.

x	y	(x, y)
-1	-1	$(-1, -1)$
0	-3	$(0, -3)$
1	-1	$(1, -1)$

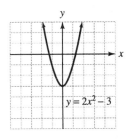

$y = 2x^2 - 3$

APPLICATIONS

41.
$$h = -16t^2 + 64t$$
$$= -16(0)^2 + 64(0)$$
$$= -16(0)$$
$$= 0 \text{ ft}$$

43.
$$h = -16t^2 + 64t$$
$$= -16(2)^2 + 64(2)$$
$$= -16(4) + 128$$
$$= -64 + 128$$
$$= 64$$

45.
$$d = 0.04v^2 + 0.9v$$
$$= 0.04(30)^2 + 0.9(30)$$
$$= 0.04(900) + 27$$
$$= 36 + 27$$
$$= 63 \text{ ft}$$

47.
$$d = 0.04v^2 + 0.9v$$
$$= 0.04(60)^2 + 0.9(60)$$
$$= 0.04(3600) + 54$$
$$= 144 + 54$$
$$= 198 \text{ ft}$$

49.

x	0	2	4	-2	-4
y	0	1	4	1	4

REVIEW

55.
$$\frac{2}{3} + \frac{4}{3} = \frac{2+4}{3}$$
$$= \frac{6}{3}$$
$$= 2$$

57.
$$\frac{1}{2} + \frac{2}{3} = \frac{1 \cdot 3}{2 \cdot 3} + \frac{2 \cdot 2}{3 \cdot 2}$$
$$= \frac{3}{6} + \frac{4}{6}$$
$$= \frac{3+4}{6}$$
$$= \frac{7}{6}$$

59.
$$\frac{36}{7} - \frac{23}{7} = \frac{36 - 23}{7}$$
$$= \frac{13}{7}$$

61.
$$\frac{5}{12} \cdot \frac{18}{5} = \frac{5 \cdot 18}{12 \cdot 5}$$
$$= \frac{\cancel{5} \cdot \cancel{6} \cdot 3}{2 \cdot \cancel{6} \cdot \cancel{5}}$$
$$= \frac{3}{2}$$

63.
$$\frac{2}{13} \div \frac{8}{13} = \frac{2}{13} \cdot \frac{13}{8}$$
$$= \frac{2 \cdot 13}{13 \cdot 8}$$
$$= \frac{\cancel{2} \cdot \cancel{13}}{\cancel{13} \cdot \cancel{2} \cdot 4}$$
$$= \frac{1}{4}$$

65.
$$x - 4 = 12$$
$$x - 4 + 4 = 12 + 4$$
$$x = 16$$

67.
$$2(x - 3) = 6$$
$$2x - 6 = 6$$
$$2x - 6 + 6 = 6 + 6$$
$$2x = 12$$
$$\frac{2x}{2} = \frac{12}{2}$$
$$x = 6$$

STUDY SET Section 6.4

VOCABULARY

1. If two algebraic terms have exactly the same variables and exponents, they are called <u>like terms</u>.

CONCEPTS

3. To add two monomials, we add the <u>coefficients</u> and keep the same <u>variables</u> and exponents.

5. $3y$ and $4y$ are like terms.
   $$3y + 4y = 7y$$

7. $3x$ and $3y$ are not like terms since x and y are different variables.

9. $3x^3$, $4x^3$, and $6x^3$ are like terms.
   $$3x^3 + 4x^3 + 6x^3 = 7x^3 + 6x^3$$
   $$= 13x^3$$

11. $-5x^2$, $13x^2$, and $7x^2$ are like terms.
   $$-5x^2 + 13x^2 + 7x^2 = 8x^2 + 7x^2$$
   $$= 15x^2$$

NOTATION

13. $(3x^2 + 2x - 5) + (2x^2 - 7x) = (3x^2 + \underline{2x^2}) + (2x - \underline{7x}) + (-5)$
   $$= \underline{5x^2} + (-5x) - 5$$
   $$= 5x^2 - 5x - 5$$

PRACTICE

15. $4y + 5y = 9y$

17. $-8t^2 - 4t^2 = -12t^2$

19. $3s^2 + 4s^2 + 7s^2 = 7s^2 + 7s^2$
$$= 14s^2$$

21. $(3x + 7) + (4x - 3) = (3x + 4x) + (7 - 3)$
$$= 7x + 4$$

23. $(2x^2 + 3) + (5x^2 - 10) = (2x^2 + 5x^2) + (3 - 10)$
$$= 7x^2 - 7$$

25. $(5x^3 - 4.2x) + (7x^3 - 10.7x) = (5x^3 + 7x^3) + (-4.2x - 10.7x)$
$$= 12x^3 - 14.9x$$

27. $(3x^2 + 2x - 4) + (5x^2 - 17) = (3x^2 + 5x^2) + (2x) + (-4 - 17)$
$$= 8x^2 + 2x - 21$$

29. $(7y^2 + 5y) + (y^2 - y - 2) = (7y^2 + y^2) + (5y - y) + (-2)$
$$= 8y^2 + 4y - 2$$

31. $(3x^2 - 3x - 2) + (3x^2 + 4x - 3) = (3x^2 + 3x^2) + (-3x + 4x) + (-2 - 3)$
$$= 6x^2 + x - 5$$

33. $(3n^2 - 5.8n + 7) + (-n^2 + 5.8n - 2) = (3n^2 - n^2) + (-5.8n + 5.8n) + (7 - 2)$
$$= 2n^2 + 5$$

35.
$$\begin{array}{r} 3x^2 + 4x + 5 \\ 2x^2 - 3x + 6 \\ \hline 5x^2 + x + 11 \end{array}$$

37.
$$\begin{array}{r} -3x^2 \quad\quad - 7 \\ -4x^2 - 5x + 6 \\ \hline -7x^2 - 5x - 1 \end{array}$$

39.
$$\begin{array}{r} -3x^2 + 4x + 25.4 \\ 5x^2 - 3x - 12.5 \\ \hline 2x^2 + x + 12.9 \end{array}$$

41. $32u^3 - 16u^3 = 16u^3$

43. $18x^5 - 11x^5 = 7x^5$

45.
$$\begin{aligned}
(4.5a + 3.7) - (2.9a - 4.3) &= (4.5a + 3.7) + [-(2.9a - 4.3)] \\
&= (4.5a + 3.7) + (-2.9a + 4.3) \\
&= (4.5a - 2.9a) + (3.7 + 4.3) \\
&= 1.6a + 8
\end{aligned}$$

47.
$$\begin{aligned}
(-8x^2 - 4) - (11x^2 + 1) &= (-8x^2 - 4) + [-(11x^2 + 1)] \\
&= (-8x^2 - 4) + (-11x^2 - 1) \\
&= (-8x^2 - 11x^2) + (-4 - 1) \\
&= -19x^2 - 5
\end{aligned}$$

49.
$$\begin{aligned}
(3x^2 - 2x - 1) - (-4x^2 + 4) &= (3x^2 - 2x - 1) + [-(-4x^2 + 4)] \\
&= (3x^2 - 2x - 1) + (4x^2 - 4) \\
&= (3x^2 + 4x^2) + (-2x) + (-1 - 4) \\
&= 7x^2 - 2x - 5
\end{aligned}$$

51.
$$\begin{aligned}
(3.7y^2 - 5) - (2y^2 - 3.1y + 4) &= (3.7y^2 - 5) + [-(2y^2 - 3.1y + 4)] \\
&= (3.7y^2 - 5) + (-2y^2 + 3.1y - 4) \\
&= (3.7y^2 - 2y^2) + (3.1y) + (-5 - 4) \\
&= 1.7y^2 + 3.1y - 9
\end{aligned}$$

53.
$$\begin{aligned}
(2b^2 + 3b - 5) - (2b^2 - 4b - 9) &= (2b^2 + 3b - 5) + [-(2b^2 - 4b - 9)] \\
&= (2b^2 + 3b - 5) + (-2b^2 + 4b + 9) \\
&= (2b^2 - 2b^2) + (3b + 4b) + (-5 + 9) \\
&= 7b + 4
\end{aligned}$$

55.
$$\begin{aligned}
(5p^2 - p + 7.1) - (4p^2 + p + 7.1) &= (5p^2 - p + 7.1) + [-(4p^2 + p + 7.1)] \\
&= (5p^2 - p + 7.1) + (-4p^2 - p - 7.1) \\
&= (5p^2 - 4p^2) + (-p - p) + (7.1 - 7.1) \\
&= p^2 - 2p
\end{aligned}$$

57.
$$\begin{array}{c}
3x^2 + 4x - 5 \\
\underline{-(-2x^2 - 2x + 3)}
\end{array} \rightarrow
\begin{array}{c}
3x^2 + 4x - 5 \\
\underline{2x^2 + 2x - 3} \\
5x^2 + 6x - 8
\end{array}$$

59.
$$\begin{array}{c}
-2x^2 - 4x + 12 \\
\underline{-(10x^2 + 9x - 24)}
\end{array} \rightarrow
\begin{array}{c}
-2x^2 - 4x + 12 \\
\underline{-10x^2 - 9x + 24} \\
-12x^2 - 13x + 36
\end{array}$$

61.
$$\begin{array}{c}
4x^3 - 3x + 10 \\
\underline{-(5x^3 - 4x - 4)}
\end{array} \rightarrow
\begin{array}{c}
4x^3 - 3x + 10 \\
\underline{-5x^3 + 4x + 4} \\
-x^3 + x + 14
\end{array}$$

APPLICATIONS

63. After 10 years, $x = 10$.

$$y = 700x + 85,000$$
$$= 700(10) + 85,000$$
$$= 7,000 + 85,000$$
$$= 92,000$$

The expected value is $92,000.

65. After 12 years, $x = 12$.

$$y = 900x + 102,000$$
$$= 900(12) + 102,000$$
$$= 10,800 + 102,000$$
$$= 112,800$$

The expected value is $112,000.

67. a. For the first house,

$$y = 700x + 85,000$$
$$= 700(15) + 85,000$$
$$= 10,500 + 85,000$$
$$= 95,500$$

For the second house,

$$y = 900x + 102,000$$
$$= 900(15) + 102,000$$
$$= 13,500 + 102,000$$
$$= 115,500$$

The value of both houses is
$95,500 + $115,500
$= $211,000$.

b. From Exercise 66,

$$y = 1,600x + 187,000$$
$$= 1,600(15) + 187,000$$
$$= 24,000 + 187,000$$
$$= 211,000$$

The value of both houses is
$211,000.

69. $y = -800x + 8,500$

71.
$$y = (-800x + 8,500) + (-1,100x + 10,200)$$
$$y = (-800x - 1,100x) + (8,500 + 10,200)$$
$$y = -1,900x + 18,700$$

REVIEW

75. $(3)(-2) + (-1)(2) = -6 + (-2)$
$$= -6 - 2$$
$$= -8$$

77. $3[(-2)^3 + (-5)] = 3(-8 - 5)$
$$= 3(-13)$$
$$= -39$$

79. $-5x + 1 = -14$
$$-5x = -15$$
$$x = 3$$

81.
$$-4\left(\frac{y}{-4} - 5\right) = -4(-2)$$
$$-4\left(\frac{y}{-4}\right) - 4(-5) = -4(-2)$$
$$y + 20 = 8$$
$$y = -12$$

83. $-3(y - 4) = 4y + 12$
$-3y + 12 = 4y + 12$
$-3y = 4y$
$0 = 7y$
$0 = y$

STUDY SET Section 6.5

VOCABULARY

1. A polynomial with one term is called a <u>monomial</u>.

3. A polynomial with <u>3</u> terms is called a trinomial.

CONCEPTS

5. To multiply two monomials, multiply the <u>numerical factors</u> and then multiply the <u>variable factors</u>.

7. To multiply two binomials, multiply each <u>term</u> of one binomial by each <u>term</u> of the other binomial and combine <u>like</u> terms.

NOTATION

9. $3x(2x - 5) = 3x \cdot \underline{2x} - 3x \cdot \underline{5}$
$= 6x^2 - 15x$

PRACTICE

11. $(3x^2)(4x^3) = (3 \cdot 4)(x^2 \cdot x^3)$
$= 12x^5$

13. $(3b^2)(-2b) = (3 \cdot -2)(b^2 \cdot b)$
$= -6b^3$

15. $(-2x^2)(3x^3) = (-2 \cdot 3)(x^2 \cdot x^3)$
$= -6x^5$

17. $\left(-\frac{2}{3}y^5\right)\left(\frac{3}{4}y^2\right) = \left(-\frac{2}{3} \cdot \frac{3}{4}\right)(y^5 \cdot y^2)$
$$= -\frac{1}{2}y^7$$

19. $3(x + 4) = 3 \cdot x + 3 \cdot 4$
$= 3x + 12$

21. $-4(t + 7) = -4 \cdot t - 4 \cdot 7$
$= -4t - 28$

23. $3x(x - 2) = 3x \cdot x - 3x \cdot 2$
$= 3x^2 - 6x$

25. $-2x^2(3x^2 - x) = -2x^2 \cdot 3x^2 - 2x^2 \cdot -x$
$$= -6x^4 + 2x^3$$

27. $2x(3x^2 + 4x - 7) = 2x \cdot 3x^2 + 2x \cdot 4x - 2x \cdot 7$
$$= 6x^3 + 8x - 14x$$

29. $-p(2p^2 - 3p + 2) = -p \cdot 2p^2 - p(-3p) - p \cdot 2$
$$= -2p^3 + 3p^2 - 2p$$

31. $3q^2(q^2 - 2q + 7) = 3q^2 \cdot q^2 - 3q^2 \cdot 2q + 3q^2 \cdot 7$
$$= 3q^4 - 6q^3 + 21q^2$$

33. $(a+4)(a+5) = (a+4)a + (a+4)(5)$
$$= a(a+4) + 5(a+4)$$
$$= a \cdot a + a \cdot 4 + 5 \cdot a + 5 \cdot 4$$
$$= a^2 + 4a + 5a + 20$$
$$= a^2 + 9a + 20$$

35. $(3x-2)(x+4) = (3x-2)x + (3x-2)4$
$$= x(3x-2) + 4(3x-2)$$
$$= x \cdot 3x - x \cdot 2 + 4 \cdot 3x - 4 \cdot 2$$
$$= 3x^2 - 2x + 12x - 8$$
$$= 3x^2 + 10x - 8$$

37. $(2a+4)(3a-5) = (2a+4)3a - (2a+4)5$
$$= 3a(2a+4) - 5(2a+4)$$
$$= 3a \cdot 2a + 3a \cdot 4 - 5 \cdot 2a - 5 \cdot 4$$
$$= 6a^2 + 12a - 10a - 20$$
$$= 6a^2 + 2a - 20$$

39. $(2x+3)^2 = (2x+3)(2x+3)$
$$= (2x+3)2x + (2x+3)3$$
$$= 2x(2x+3) + 3(2x+3)$$
$$= 2x \cdot 2x + 2x \cdot 3 + 3 \cdot 2x + 3 \cdot 3$$
$$= 4x^2 + 6x + 6x + 9$$
$$= 4x^2 + 12x + 9$$

41. $(2x-3)^2 = (2x-3)(2x-3)$
$$= (2x-3)2x - (2x-3)3$$
$$= 2x(2x-3) - 3(2x-3)$$
$$= 2x \cdot 2x - 2x \cdot 3 - 3 \cdot 2x - 3(-3)$$
$$= 4x^2 - 6x - 6x + 9$$
$$= 4x^2 - 12x + 9$$

43. $(5t+1)^2 = (5t+1)(5t+1)$
$$= (5t+1)5t + (5t+1)1$$
$$= 5t(5t+1) + 1(5t+1)$$
$$= 5t \cdot 5t + 5t \cdot 1 + 5t + 1$$
$$= 25t^2 + 5t + 5t + 1$$
$$= 25t^2 + 10t + 1$$

45. $(2x+1)(3x^2-2x+1) = (2x+1)3x^2 - (2x+1)2x + (2x+1)1$
$$= 3x^2(2x+1) - 2x(2x+1) + 1(2x+1)$$
$$= 3x^2 \cdot 2x + 3x^2 \cdot 1 - 2x \cdot 2x - 2x \cdot 1 + 2x + 1$$
$$= 6x^3 + 3x^2 - 4x^2 - 2x + 2x + 1$$
$$= 6x^3 - x^2 + 1$$

47. $(x-1)(x^2+x+1) = (x-1)x^2 + (x-1)x + (x-1)1$
$$= x^2(x-1) + x(x-1) + 1(x-1)$$
$$= x^2 \cdot x - x^2 \cdot 1 + x \cdot x - x \cdot 1 + x - 1$$
$$= x^3 - x^2 + x^2 - x + x - 1$$
$$= x^3 - 1$$

49. $(x+2)(x^2-3x+1) = (x+2)x^2 - (x+2)3x + (x+2)1$
$$= x^2(x+2) - 3x(x+2) + 1(x+2)$$
$$= x^2 \cdot x + x^2 \cdot 2 - 3x \cdot x - 3x \cdot 2 + x + 2$$
$$= x^3 + 2x^2 - 3x^2 - 6x + x + 2$$
$$= x^3 - x^2 - 5x + 2$$

51.
$$\begin{array}{r} 4x+3 \\ \times \quad x+2 \\ \hline 4x^2 + 3x \\ + 8x + 6 \\ \hline 4x^2 + 11x + 6 \end{array}$$

53.
$$\begin{array}{r} 4x-2 \\ \times \quad 3x+5 \\ \hline 12x^2 - 6x \\ + 20x - 10 \\ \hline 12x^2 + 14x - 10 \end{array}$$

55.

$$x^2 - x + 1$$
$$\underline{\times \quad x + 1}$$
$$x^3 - x^2 + x$$
$$\underline{\quad + x^2 - x + 1}$$
$$x^3 \qquad\qquad + 1$$

APPLICATIONS

57.
$$\begin{aligned}
(x - 2)(x + 2) &= (x - 2)x + (x - 2)2 \\
&= x(x - 2) + 2(x - 2) \\
&= x \bullet x - x \bullet 2 + 2 \bullet x - 2 \bullet 2 \\
&= x^2 - 2x + 2x - 4 \\
&= x^2 - 4
\end{aligned}$$

The area is $(x^2 - 4)$ ft^2.

59. $R = (\text{price})(\text{number sold})$

$$R = \left(-\frac{x}{100} + 30 \right) x$$

$$R = x \left(-\frac{x}{100} + 30 \right)$$

$$R = x \left(-\frac{x}{100} \right) + x \bullet 30$$

$$R = -\frac{x^2}{100} + 30x$$

REVIEW

63. 7,507

65.
$$\begin{aligned}
P &= 2l + 2w \\
&= 2(757) + 2(327) \\
&= 1,514 + 654 \\
&= 2,168
\end{aligned}$$
The perimeter is 2,168 in.

67.
$$\begin{aligned}
3 \bullet 4^2 - 2^3 &= 3 \bullet 16 - 8 \\
&= 48 - 8 \\
&= 40
\end{aligned}$$

69. $x^3x^2 = x^{3+2}$
$$= x^5$$

71. $(x^2x^3)^3 = (x^{2+3})^3$
$$= (x^5)^3$$
$$= x^{5 \cdot 3}$$
$$= x^{15}$$

KEY CONCEPT Graphing

1.

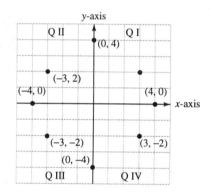

3. (3, 2)

5. (0, 0)

7.

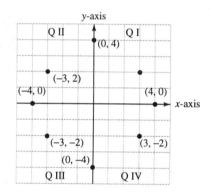

CHAPTER 6 REVIEW

1. a.
$$2x + 5y = -11$$
$$2(2) + 5(-3) \overset{?}{=} -11$$
$$4 - 15 \overset{?}{=} -11$$
$$-11 = -11$$

Yes, $x = 2$ and $y = -3$ is a solution of $2x + 5y = -11$.

1. b.
$$3x - 5y = 19$$
$$3(-3) - 5(2) \overset{?}{=} 19$$
$$-9 - 10 \overset{?}{=} -11$$
$$-19 \neq 19$$

No, $x = -3$ and $y = 2$ is not a solution of $3x - 5y = 19$.

3. a.

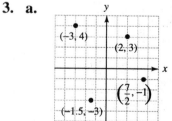

3. b. $A(4, 3)$, $B(-3, 3)$, $C(-4, 0)$,

$D\left(-\dfrac{3}{2}, -\dfrac{7}{2}\right)$, $E\left(\dfrac{5}{2}, -\dfrac{3}{2}\right)$

5. a.

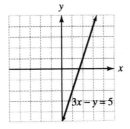

$3x - y = 5$

b.

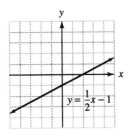

$y = \dfrac{1}{2}x - 1$

7. a. $3x^2 + 4x - 5$ has three terms, so it is a trinomial.

b. $3t^2$ has one term, so it is a monomial.

c. $2x^2 - 1$ has two terms, so it is a binomial.

9. a.
$$3x^2 - 2x - 1 = 3(2)^2 - 2(2) - 1$$
$$= 3(4) - 4 - 1$$
$$= 12 - 4 - 1$$
$$= 8 - 1$$
$$= 7$$

b.
$$2t^2 + t - 2 = 2(-3)^2 + (-3) - 2$$
$$= 2(9) - 3 - 2$$
$$= 18 - 3 - 2$$
$$= 15 - 2$$
$$= 13$$

11. a. $3x^3 + 2x^3 = 5x^3$

b.
$$\dfrac{1}{2}p^2 + \dfrac{5}{2}p^2 + \dfrac{7}{2}p^2 = \dfrac{6}{2}p^2 + \dfrac{7}{2}p^2$$
$$= \dfrac{13}{2}p^2$$

13. a.
$$\begin{array}{r} 5x - 2 \\ 3x + 5 \\ \hline 8x + 3 \end{array}$$

b.
$$\begin{array}{r} 3x^2 - 2x + 7 \\ -5x^2 + 3x - 5 \\ \hline -2x^2 + x + 2 \end{array}$$

15. a.
$$(2.5x + 4) - (1.4x + 12) = (2.5x + 4) + [-(1.4x + 12)]$$
$$= (2.5x + 4) + (-1.4x - 12)$$
$$= (2.5x - 1.4x) + (4 - 12)$$
$$= 1.1x - 8$$

b.
$$(3z^2 - z + 4) - (2z^2 + 3z - 2) = (3z^2 - z + 4) + [-(2z^2 + 3z - 2)]$$
$$= (3z^2 - z + 4) + (-2z^2 - 3z + 2)$$
$$= (3z^2 - 2z^2) + (-z - 3z) + (4 + 2)$$
$$= z^2 - 4z + 6$$

17. a.
$$3x^2 \bullet 5x^3 = (3 \bullet 5)(x^2 \bullet x^3)$$
$$= 15x^5$$

b.
$$(3z^2)(-2z^2) = (3 \bullet -2)(z^2 \bullet z^2)$$
$$= -6z^4$$

19. a.
$$(2x - 1)(3x + 2) = (2x - 1)3x + (2x - 1)2$$
$$= 3x(2x - 1) + 2(2x - 1)$$
$$= 3x \bullet 2x - 3x \bullet 1 + 2 \bullet 2x - 2 \bullet 1$$
$$= 6x^2 - 3x + 4x - 2$$
$$= 6x^2 + x - 2$$

b.
$$(5t + 4)(7t - 6) = (5t + 4)7t - (5t + 4)6$$
$$= 7t(5t + 4) - 6(5t + 4)$$
$$= 7t \bullet 5t + 7t \bullet 4 - 6 \bullet 5t - 6 \bullet 4$$
$$= 35t^2 + 28t - 30t - 24$$
$$= 35t^2 - 2t - 24$$

21. a.
$$(3x + 2)(2x^2 - x + 1) = (3x + 2)2x^2 - (2x + 2)x + (3x + 2)1$$
$$= 2x^2(3x + 2) - x(3x + 2) + 1(3x + 2)$$
$$= 2x^2 \bullet 3x + 2x^2 \bullet 2 - x \bullet 3x - x \bullet 2 + 3x + 2$$
$$= 6x^3 + 4x^2 - 3x^2 - 2x + 3x + 2$$
$$= 6x^3 + x^2 + x + 2$$

b.
$$(2r - 3)(3r^2 + 2r - 3) = (2r - 3)3r^2 + (2r - 3)2r - (2r - 3)3$$
$$= 3r^2(2r - 3) + 2r(2r - 3) - 3(2r - 3)$$
$$= 3r^2(2r) - 3r^2(3) + 2r(2r) - 2r(3) - 3(2r) - 3(-3)$$
$$= 6r^3 - 9r^2 + 4r^2 - 6r - 6r + 9$$
$$= 6r^3 - 5r^2 - 12r + 9$$

CHAPTER 6 TEST

1.
$$4x + 5y = 6$$
$$4(-1) + 5(2) \stackrel{?}{=} 6$$
$$-4 + 10 \stackrel{?}{=} 6$$
$$6 = 6$$
Yes, $x = -1$, $y = 2$ is a solution of $4x + 5y = 6$.

3. $x = 0$:
$$x - 2y = 4$$
$$0 - 2y = 4$$
$$-2y = 4$$
$$y = -2$$
$(0, -2)$

$y = 0$:
$$x - 2y = 4$$
$$x - 2(0) = 4$$
$$x - 0 = 4$$
$$x = 4$$
$(4, 0)$

$x = 2$:
$$x - 2y = 4$$
$$2 - 2y = 4$$
$$-2y = 2$$
$$y = -1$$
$(2, -1)$

5.

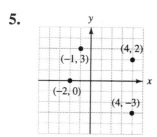

7.

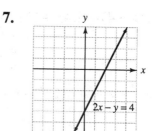

9.

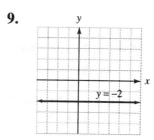

11. $5x^2 + 4x$ has two terms, so it is a binomial.

13. $3t^4$ has degree 4, $-2t^3$ has degree 3, $5t^6$ has degree 6, and $-t$ has degree 1, so $3t^4 - 2t^3 + 5t^6 - t$ has degree 6.

15.
$$3x^2 - 2x + 4 = 3(3)^2 - 2(3) + 4$$
$$= 3(9) - 6 + 4$$
$$= 27 - 6 + 4$$
$$= 21 + 4$$
$$= 25$$

17.

x	y	(x, y)
-3	1	$(-3, 1)$
0	-2	$(0, -2)$
3	1	$(3, 1)$

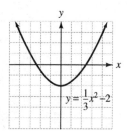

19. $(3x^2 + 2x) + (2x^2 - 5x + 4) = (3x^2 + 2x^2) + (2x - 5x) + (4)$
$$= 5x^2 - 3x + 4$$

21. $(2.1p^2 - 2p - 2) - (3.3p^2 - 5p - 2) = (2.1p^2 - 2p - 2) + [-(3.3p^2 - 5p - 2)]$
$$= (2.1p^2 - 2p - 2) + (-3.3p^2 + 5p + 2)$$
$$= (2.1p^2 - 3.3p^2) + (-2p + 5p) + (-2 + 2)$$
$$= -1.2p^2 + 3p$$

23. $(-2x^3)(4x^2) = (-2 \bullet 4)(x^3 \bullet x^2)$
$$= -8x^5$$

25. $(2x - 5)(3x + 4) = (2x - 5)3x + (2x - 5)4$
$$= 3x(2x - 5) + 4(2x - 5)$$
$$= 3x \bullet 2x - 3x \bullet 5 + 4 \bullet 2x - 4 \bullet 5$$
$$= 6x^2 - 15x + 8x - 20$$
$$= 6x^2 - 7x - 20$$

CHAPTERS 1–6 CUMULATIVE REVIEW

1. 6,246,000

3. $P = 2l + 2w$
$$= 2(8) + 2(3)$$
$$= 16 + 6$$
$$= 22$$
22 m

5. $A = l \bullet w$
$$= (8)(3)$$
$$= 24$$
24 m^2

7. $120 = 2^3 \bullet 3 \bullet 5$

9. $3x + 2 = 3(-2) + 2$
$$= -6 + 2$$
$$= -4$$

11. $5x - 11x = -6x$

13. $4x + 3 = 11$
$4x = 8$
$x = 2$
Check:
$4x + 3 = 11$
$4(2) + 3 \stackrel{?}{=} 11$
$8 + 3 \stackrel{?}{=} 11$
$11 = 11$

15. $\dfrac{t}{3} + 2 = -4$

$\dfrac{t}{3} = -6$

$t = -6 \cdot 3$

$t = -18$
Check:

$\dfrac{t}{3} + 2 = -4$

$\dfrac{-18}{3} + 2 \stackrel{?}{=} -4$

$-6 + 2 \stackrel{?}{=} -4$

$-4 = -4$

17. $2y + 7 = 2 - (4y + 7)$
$2y + 7 = 2 - 4y - 7$
$2y + 7 = -4y - 5$
$6y + 7 = -5$
$6y = -12$
$y = -2$
Check:
$2y + 7 = 2 - (4y + 7)$
$2(-2) + 7 \stackrel{?}{=} 2 - [4(-2) + 7]$
$-4 + 7 \stackrel{?}{=} 2 - (-8 + 7)$
$3 \stackrel{?}{=} 2 - (-1)$
$3 = 3$

19. $p^4 p^5 = p^{4+5}$
$= p^9$

21. $(p^3 q^2)(p^3 q^4) = (p^3 \cdot p^3)(q^2 \cdot q^4)$
$= p^{3+3} q^{2+4}$
$= p^6 q^6$

23. $\dfrac{5a^4}{10b^3} \cdot \dfrac{2b^2}{a^3} = \dfrac{(5a^4)(2b^2)}{(10b^3)(a^3)}$

$= \dfrac{10a^4 b^2}{10a^3 b^3}$

$= \dfrac{10a^3 \cdot a \cdot b^2}{10a^3 \cdot b^2 \cdot b}$

$= \dfrac{a}{b}$

25. $\dfrac{x}{2} + \dfrac{3}{5} = \dfrac{x \cdot 5}{2 \cdot 5} + \dfrac{3 \cdot 2}{5 \cdot 2}$

$= \dfrac{5x}{10} + \dfrac{6}{10}$

$= \dfrac{5x + 6}{10}$

27.
$$x - \frac{1}{4} = \frac{4}{5}$$

$$x = \frac{4}{5} + \frac{1}{4}$$

$$x = \frac{4 \cdot 4}{5 \cdot 4} + \frac{1 \cdot 5}{4 \cdot 5}$$

$$x = \frac{16}{20} + \frac{5}{20}$$

$$x = \frac{16 + 5}{20}$$

$$x = \frac{21}{20}$$

Check:

$$x - \frac{1}{4} = \frac{4}{5}$$

$$\frac{21}{20} - \frac{1}{4} \stackrel{?}{=} \frac{4}{5}$$

$$\frac{21}{20} - \frac{1 \cdot 5}{4 \cdot 5} \stackrel{?}{=} \frac{4}{5}$$

$$\frac{21}{20} - \frac{5}{20} \stackrel{?}{=} \frac{4}{5}$$

$$\frac{21 - 5}{20} \stackrel{?}{=} \frac{4}{5}$$

$$\frac{16}{20} \stackrel{?}{=} \frac{4}{5}$$

$$\frac{4 \cdot 4}{4 \cdot 5} \stackrel{?}{=} \frac{4}{5}$$

$$\frac{4}{5} = \frac{4}{5}$$

29. 57.6

31. $29.703 + 321.35 = 29.703 + 321.350$
$$= 351.053$$

33. $789 \times 27 = 21{,}303$ so
$7.89 \times 0.27 = 2.1303$

35.
$$\begin{array}{r} 0.6 \\ 5{\overline{)3.0}} \\ \underline{30} \\ 0 \end{array}$$

$$\frac{3}{5} = 0.6$$

37.
$$\begin{array}{r} 0.625 \\ 8{\overline{)5.000}} \\ \underline{4.8} \\ 20 \\ \underline{16} \\ 40 \\ \underline{40} \\ 0 \end{array}$$

Thus, $\frac{5}{8} = 0.625$ and

$$5\frac{5}{8} = 5 + \frac{5}{8} = 5 + 0.625 = 5.625.$$

Rounded to the nearest tenth,

$$5\frac{5}{8} \approx 5.6.$$

39.
$$3.2x = 74.46 - 1.9x$$
$$3.2x + 1.9x = 74.46$$
$$5.1x = 74.46$$
$$x = 14.6$$

Check:
$$3.2x = 74.46 - 1.9x$$
$$3.2(14.6) \stackrel{?}{=} 74.46 - 1.9(14.6)$$
$$46.72 \stackrel{?}{=} 74.46 - 27.74$$
$$46.72 = 46.72$$

41. $-2(x - 2.1) = -2.4$
$$-2x + 4.2 = -2.4$$
$$-2x = -6.6$$
$$x = 3.3$$

Check:
$$-2(x - 2.1) = -2.4$$
$$-2(3.3 - 2.1) \stackrel{?}{=} -2.4$$
$$-2(1.2) \stackrel{?}{=} -2.4$$
$$-2.4 = -2.4$$

43. Let x = the number of signatures.
$$20 + 0.05x = 60$$
$$0.05x = 40$$
$$x = 800$$
She needs to collect 800 signatures.

45. $\sqrt{121} = \sqrt{11 \cdot 11}$
$$= 11$$

47. $\sqrt{0.25} = \sqrt{0.5 \cdot 0.5}$
$$= 0.5$$

49. $4x - 5y = -23$
$$4(-2) - 5(3) \stackrel{?}{=} -23$$
$$-8 - 15 \stackrel{?}{=} -23$$
$$-23 = -23$$
Yes, $(-2, 3)$ is a solution of
$4x - 5y = -23$.

51. $(3x^2 - 5x) - (2x^2 + x - 3) = (3x^2 - 5x) + [-(2x^2 + x - 3)]$
$$= (3x^2 - 5x) + (-2x^2 - x + 3)$$
$$= (3x^2 - 2x^2) + (-5x - x) + (3)$$
$$= x^2 - 6x + 3$$

VOCABULARY

1. <u>Percent</u> means per one hundred.

CONCEPTS

3. To change a percent to a fraction, drop the percent sign and write the given number over <u>100</u>.

5. To change a decimal to a percent, move the decimal point two places to the <u>right</u> and insert a % sign.

7. **a.** 84 of the 100 squares are shaded. This is 0.84, 84%, or
$$\frac{84}{100} = \frac{4 \cdot 21}{4 \cdot 25} = \frac{21}{25}.$$

    b. 16 of the 100 squares are not shaded. This is 16%.

PRACTICE

9. $17\% = \dfrac{17}{100}$

11. $5\% = \dfrac{5}{100}$

$$= \frac{1}{20}$$

13. $60\% = \dfrac{60}{100}$

$$= \frac{3}{5}$$

15. $125\% = \dfrac{125}{100}$

$$= \frac{5}{4}$$

17. $\dfrac{2}{3}\% = \dfrac{\frac{2}{3}}{100}$

$$= \frac{2}{3} \div 100$$

$$= \frac{2}{3} \cdot \frac{1}{100}$$

$$= \frac{2 \cdot 1}{3 \cdot 100}$$

$$= \frac{1}{3 \cdot 50}$$

$$= \frac{1}{150}$$

19. $5\dfrac{1}{4}\% = \dfrac{5\frac{1}{4}}{100}$

$$= 5\frac{1}{4} \div 100$$

$$= \frac{21}{4} \cdot \frac{1}{100}$$

$$= \frac{21 \cdot 1}{4 \cdot 100}$$

$$= \frac{21}{400}$$

21. $0.6\% = \dfrac{0.6}{100}$

$$= \frac{0.6 \cdot 10}{100 \cdot 10}$$

$$= \frac{6}{1,000}$$

$$= \frac{3}{500}$$

23. $1.9\% = \dfrac{1.9}{100}$

$$= \frac{1.9 \cdot 10}{100 \cdot 10}$$

25. $19\% = 19.0\%$
$= .190$
$= 0.19$

27. $6\% = 6.0\%$
$= .060$
$= 0.06$

29. $40.8\% = .408$
$= 0.408$

31. $250\% = 250.0\%$
$= 2.500$
$= 2.5$

33. $0.79\% = .0079$
$= 0.0079$

35. $\dfrac{1}{4}\% = 0.25\%$
$= .0025$
$= 0.0025$

37. $0.93 = 93\%$

39. $0.612 = 61.2\%$

41. $0.0314 = 3.14\%$

43. $8.43 = 843\%$

45. $50 = 5{,}000\%$

47. $9.1 = 910\%$

49. $\dfrac{17}{100} = 17\%$

51.
$$
\begin{array}{r}
0.16 \\
25\overline{)4.00} \\
2\,5 \\
\hline
1\,50 \\
1\,50 \\
\hline
0
\end{array}
$$
$\dfrac{4}{25} = 0.16$
$= 16\%$

53.
$$
\begin{array}{r}
0.4 \\
5\overline{)2.0} \\
2\,0 \\
\hline
0
\end{array}
$$
$\dfrac{2}{5} = 0.4$
$= 40\%$

55.
$$
\begin{array}{r}
1.05 \\
20\overline{)21.00} \\
20 \\
\hline
100 \\
100 \\
\hline
0
\end{array}
$$
$\dfrac{21}{20} = 1.05$
$= 105\%$

57.
$$
\begin{array}{r}
0.625 \\
8\overline{)5.000} \\
4\,8 \\
\hline
20 \\
16 \\
\hline
40 \\
40 \\
\hline
0
\end{array}
$$
$\dfrac{5}{8} = 0.625$
$= 62.5\%$

59.
$$
\begin{array}{r}
0.1875 \\
16\overline{)3.0000} \\
1\,6 \\
\hline
1\,40 \\
1\,28 \\
\hline
120 \\
112 \\
\hline
80 \\
80 \\
\hline
0
\end{array}
$$
$\dfrac{3}{16} = 0.1875$
$= 18.75\%$

61.

$$\begin{array}{r} 0.666 \\ 3\overline{\smash{)}2.000} \\ \underline{1\ 8} \\ 20 \\ \underline{18} \\ 20 \\ \underline{18} \\ 2 \end{array}$$

$$\frac{2}{3} = 0.\overline{6}$$
$$= 0.66\overline{6}$$
$$= 66.\overline{6}\%$$
$$= 66\frac{2}{3}\%$$

63.

$$\begin{array}{r} 0.833 \\ 6\overline{\smash{)}5.000} \\ \underline{4\ 8} \\ 20 \\ \underline{18} \\ 20 \\ \underline{18} \\ 2 \end{array}$$

$$\frac{5}{6} = 0.8\overline{3}$$
$$= 0.83\overline{3}$$
$$= 83.\overline{3}\%$$
$$= 83\frac{1}{3}\%$$

65.

$$\begin{array}{r} 0.11 \\ 9\overline{\smash{)}1.00} \\ \underline{9} \\ 10 \\ \underline{9} \\ 1 \end{array}$$

$$\frac{1}{9} = 0.\overline{1}$$
$$= 0.1111\overline{1}$$
$$= 11.11\overline{1}\%$$
$$\approx 11.11\%$$

67.

$$\begin{array}{r} 0.55 \\ 9\overline{\smash{)}5.00} \\ \underline{4\ 5} \\ 50 \\ \underline{45} \\ 5 \end{array}$$

$$\frac{5}{9} = 0.\overline{5}$$
$$= 0.5555\overline{5}$$
$$= 55.55\overline{5}\%$$
$$\approx 55.56\%$$

APPLICATIONS

69. a. 15 of the 184 member nations are on the Security Council.

This is $\dfrac{15}{184}$ of the members.

b.

$$\begin{array}{r} 0.081 \\ 184\overline{\smash{)}15.000} \\ \underline{14\ 72} \\ 280 \\ \underline{184} \\ 96 \end{array}$$

$$\frac{15}{184} \approx 0.081$$
$$\approx 8\%$$

71. a. $\dfrac{36}{88} = \dfrac{4 \cdot 9}{4 \cdot 22}$

$$= \frac{9}{22}$$

71. b.

$$
\begin{array}{r}
0.409 \\
22\overline{)9.000} \\
\underline{8\,8} \\
200 \\
\underline{198} \\
2
\end{array}
$$

$$\frac{9}{22} \approx 0.409$$

$$\approx 41\%$$

73. a. There are $7 + 12 + 5 + 1 + 4$ $= 29$ total vertebrae, 5 of which are lumbar vertebrae. $\dfrac{5}{29}$ of the vertebrae are lumbar.

b.

$$
\begin{array}{r}
0.172 \\
29\overline{)5.000} \\
\underline{2\,9} \\
210 \\
\underline{2\,03} \\
70 \\
\underline{58} \\
12
\end{array}
$$

$$\frac{5}{29} \approx 0.17$$

$$\approx 17\%$$

17% of the vertebrae are lumbar.

73. c. 7 of the 29 vertebrae are cervical.

$$
\begin{array}{r}
0.241 \\
29\overline{)7.000} \\
\underline{5\,8} \\
120 \\
\underline{116} \\
40 \\
\underline{29} \\
11
\end{array}
$$

$$\frac{7}{29} \approx 0.241$$

$$\approx 24\%$$

24% of the vertebrae are cervical.

75. $5\% = \dfrac{5}{100}$

The road rises 5 feet over a 100-ft run.

77. $99\dfrac{44}{100}\% = 99.44\%$

$$= .9944$$
$$= 0.9944$$

79. The winning percentage is presented as the decimal .896.

$.896 = 89.6\%$

81. $\dfrac{1}{365} \approx 0.0027$

$$\approx 0.27\%$$

REVIEW

87.

$$-\frac{2}{3}x = -6$$

$$3\left(-\frac{2}{3}x\right) = 3(-6)$$

$$-2x = -18$$

$$x = 9$$

89. $x = 2$:
$y = 2x + 3$
$y = 2(2) + 3$
$y = 4 + 3$
$y = 7$
$(2, 7)$

$x = 4$:
$y = 2x + 3$
$y = 2(4) + 3$
$y = 8 + 3$
$y = 11$
$(4, 11)$

$x = 0$:
$y = 2x + 3$
$y = 2(0) + 3$
$y = 0 + 3$
$y = 3$
$(0, 3)$

91.
$$(x+1)(x+2) = (x+1)x + (x+1)2$$
$$= x(x+1) + 2(x+1)$$
$$= x \bullet x + x \bullet 1 + 2 \bullet x + 2 \bullet 1$$
$$= x^2 + x + 2x + 2$$
$$= x^2 + 3x + 2$$

STUDY SET Section 7.2

VOCABULARY

1. What number is 10% of 50?
$\underline{x = 0.10 \bullet 50}$

3. 48 is what percent of 47?
$\underline{48 = x \bullet 47}$

5. A <u>circle</u> graph can be used to show the division of a whole quantity into its component parts.

CONCEPTS

7. a. $12\% = 12.0\%$
$= .120$
$= 0.12$

b. $5.6\% = .056$
$= 0.056$

c. $125\% = 125.0\%$
$= 1.250$
$= 1.25$

7. d. $\dfrac{1}{4}\% = 0.25\%$
$$= .0025$$
$$= 0.0025$$

9.   Since 120% is more than 100%, 120% of 55 is more than 100% of 55 which is 55.

11.   Part-time employees are $100\% - 81\% = 19\%$ of the company's employees.

13. $100\% \cdot 25 = 1 \cdot 25$
$$= 25$$
25 is 100% of 25.

15. $200\% \cdot 25 = 2 \cdot 25$
$$= 50$$
50 is 200% of 25.

17. $36\% \cdot 250 = 0.36 \cdot 250$
$$= 90$$
90 is 36% of 250.

19. $16 = x \cdot 20$
$$\dfrac{16}{20} = x$$
$$0.8 = x$$
$80\% = x$
16 is 80% of 20.

21. $7.8 = 12\% \cdot x$
$$7.8 = 0.12x$$
$$\dfrac{7.8}{0.12} = x$$
$$65 = x$$
7.8 is 12% of 65.

23. $0.8\% \cdot 12 = 0.008 \cdot 12$
$$= 0.096$$
0.096 is 0.8% of 12.

25. $0.5 = x \cdot 40$
$$\dfrac{0.5}{40} = x$$
$$0.0125 = x$$
$1.25\% = x$
0.5 is 1.25% of 40.

27. $3.3 = 7.5\% \cdot x$
$$3.3 = 0.075x$$
$$\dfrac{3.3}{0.075} = x$$
$$44 = x$$
3.3 is 7.5% of 44.

29.   What number is $7\dfrac{1}{4}\%$ of 600?
$$x = 7\dfrac{1}{4}\% \cdot 600$$
$$x = 7.25\% \cdot 600$$
$$x = 0.0725 \cdot 600$$
$$x = 43.5$$
43.5 is $7\dfrac{1}{4}\%$ of 600.

31.   What number is 102% of 105?
$x = 102\% \cdot 105$
$x = 1.02 \cdot 105$
$x = 107.1$
107.1 is 102% of 105.

33. 33 is $33\frac{1}{3}\%$ of what number?

$$33 = 33\frac{1}{3}\% \cdot x$$

$$33 = \frac{1}{3} \cdot x$$

$$3 \cdot 33 = x$$

$$99 = x$$

33 is $33\frac{1}{3}\%$ of 99.

35. 5.7 is $9\frac{1}{2}\%$ of what number?

$$5.7 = 9\frac{1}{2}\% \cdot x$$

$$5.7 = 9.5\% \cdot x$$

$$5.7 = 0.095x$$

$$\frac{5.7}{0.095} = x$$

$$60 = x$$

5.7 is $9\frac{1}{2}\%$ of 60.

37. $2{,}500$ is what percent of $8{,}000$?

$$2{,}500 = x \cdot 8{,}000$$

$$\frac{2{,}500}{8{,}000} = x$$

$$0.3125 = x$$

$$31.25\% = x$$

$2{,}500$ is 31.25% of 8.000.

39.

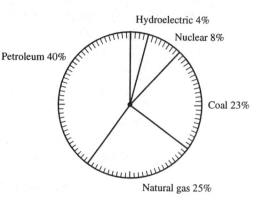

Hydroelectric 4%

Nuclear 8%

Petroleum 40%

Coal 23%

Natural gas 25%

APPLICATIONS

41. 84 is 70% of what number?

$$84 = 70\% \cdot x$$

$$84 = 0.7x$$

$$\frac{84}{0.7} = x$$

$$120 = x$$

The day care center could enroll at most 120 children.

43. What number is 35% of $1{,}500$ billion?

$$x = 35\% \cdot 1{,}500$$

$$x = 0.35 \cdot 1{,}500$$

$$x = 525$$

$\$525$ billion was spent on Social Security, Medicare, and other retirement programs.

45. 6 is 25% of what number?

$$6 = 25\% \cdot x$$

$$6 = 0.25x$$

$$\frac{6}{0.25} = x$$

$$24 = x$$

The large bottle contains 24 ounces.

47. 28 is what percent of 40?

$$28 = x \cdot 40$$

$$\frac{28}{40} = x$$

$$0.7 = x$$

$$70\% = x$$

Since 70% is passing, the man passed the test.

49. What number is 50% of 60?

$$x = 50\% \cdot 60$$

$$x = 0.5 \cdot 60$$

$$x = 30$$

What number is 30% of 40?

$$x = 30\% \cdot 40$$

$$x = 0.3 \cdot 40$$

$$x = 12$$

Milliliters of solution in beaker	% sulfuric acid	Milliliters of sulfuric acid in beaker
60	50%	30
40	30%	12

51. What number is 180% of 1.5?

$$x = 180\% \cdot 1.5$$

$$x = 1.8 \cdot 1.5$$

$$x = 2.7$$

The type on the copy will be 2.7 inches in height.

53. $200 is what percent of $4,000?

$$200 = x \cdot 4,000$$

$$\frac{200}{4,000} = x$$

$$0.05 = x$$

$$5\% = x$$

The driver paid 5% of the repair cost.

55. 39 is what percent of 317?

$$39 = x \cdot 317$$

$$\frac{39}{317} = x$$

$$0.12 \approx x$$

$$12\% \approx x$$

Approximately 12% of his business was repeat business.

REVIEW

61.
$$2.78 + 6 + 9.09 + 0.3 = 8.78 + 9.09 + 0.3$$
$$= 17.87 + 0.3$$
$$= 18.17$$

63. $5 - 4.9 = 0.1$
$5.001 - 5 = 0.001$
Since $0.001 < 0.1$, 5.001 is closer to 5 than 4.9.

65. $34.5464 \cdot 1{,}000 = 34{,}546.4$

67. $0.4x + 1.2 = -7.8$
$$0.4x = -9$$
$$x = -\frac{9}{0.4}$$
$$x = -22.5$$

STUDY SET Section 7.3

VOCABULARY

1. Some salespeople are paid on <u>commission</u>. It is based on a percent of the total dollar amount of the goods or services they sell.

3. The difference between the original price and the sale price of an item is called the <u>discount</u>.

CONCEPTS

5. If the original membership is represented by x, a 100% increase means that the new number of members is
$$x + 100\% \cdot x = x + 1 \cdot x$$
$$= x + x$$
$$= 2x$$
The number of members has doubled.

APPLICATIONS

7. What number is 3% of 25?
$x = 3\% \cdot 25?$
$x = 0.03 \cdot 25$
$x = 0.75$
The sales tax is $0.75.

9. $10.32 is what percent of $129?
$10.32 = x \cdot 129$
$$\frac{10.32}{129} = x$$
$$0.08 = x$$
$$8\% = x$$
The room tax rate is 8%.

11. The subtotal is

$$\$8.97 + \$9.87 + \$28.50 = \$18.84 + \$28.50$$
$$= \$47.34$$

For the tax, we find the number that is 6.00% of $47.34.

$x = 6.00\% \cdot 47.34$

$x = 0.06 \cdot 47.34$

$x = 2.8404$

The sales tax is $2.84. The total is $47.34 + $2.84 = $50.18.

13. 1% of $15,000 is what number?

$$1\% \cdot 15,000 = 0.01 \cdot 15,000$$
$$= 150$$

An additional $150 will be collected.

15. Federal withholding:
$28.80 is what percent of $360.00?

$28.8 = x \cdot 360$

$$\frac{28.8}{360} = x$$

$0.08 = x$

$8\% = x$

The tax rate for federal withholding is 8%.

Workmen's compensation:
$4.32 is what percent of $360.00?

$4.32 = x \cdot 360$

$$\frac{4.32}{360} = x$$

$0.012 = x$

$1.2\% = x$

The tax rate for workmen's compensation is 1.2%.

17. The increase in enrollment is $12,900 - 12,000 = 900$. 900 is what percent of 12,000?

$900 = x \cdot 12,000$

$$\frac{900}{12,000} = x$$

$0.075 = x$

$7.5\% = x$

The percent of increase in enrollment is 7.5%.

19. The percent of decrease in calories is 36%. 36% of 150 is what number?

$36\% \cdot 150 = x$

$0.36 \cdot 150 = x$

$54 = x$

The number of calories was reduced by 54, so the new chips have $150 - 54 = 96$ calories per serving.

21. The percent of increase in salary is 2.4%. 2.4% of $32,000 is what number?

$2.4\% \cdot 32,000 = x$

$0.024 \cdot 32,000 = x$

$768 = x$

The amount of her raise is $768, so her new salary is
$32,000 + $768 = $32,768.

23. The amount of the decrease was $400 – $360 = $40. $40 is what percent of $400?

$$40 = x \cdot 400$$

$$\frac{40}{400} = x$$

$$0.1 = x$$

$$10 = x$$

The percent of decrease in the premium was 10%.

25. The increase in the shoreline was $7.6 - 5.8 = 1.8$ miles. 1.8 miles is what percent of 5.8 miles?

$$1.8 = x \cdot 5.8$$

$$\frac{1.8}{5.8} = x$$

$$0.31 \approx x$$

$$31\% \approx x$$

The percent of increase in the shoreline was approximately 31%.

27. 6% of $98,500 is what number?

$$6\% \cdot 98,500 = x$$
$$0.06 \cdot 98,500 = x$$
$$5,910 = x$$

Thus, the real estate agent split $5,910 with the other agent. Each agent received half of $5,910.

$$\frac{1}{2} \cdot 5,910 = 2,955$$

Each person received $2,955.

29. $37,500 is what percent of $2,500,000?

$$37,500 = x \cdot 2,500,000$$

$$\frac{37,500}{2,500,000} = x$$

$$0.015 = x$$

$$1.5\% = x$$

The agent charged 1.5% for his services.

31. If 6,000 cars are parked, the revenue of the parking concession will be $6,000 \cdot \$6 = \$36,000$. The promoter receives $33\frac{1}{3}\%$ of this.

$33\frac{1}{3}\%$ of $36,000 is what number?

$33\frac{1}{3}\%$ is $\frac{1}{3}$.

$$33\frac{1}{3}\% \cdot 36,000 = x$$

$$\frac{1}{3} \cdot 36,000 = x$$

$$12,000 = x$$

The promoter would receive $12,000.

33. The regular price of the watch is $29.95 + $10 = $39.95. The discount amount is $10. $10 is what percent of $39.95?

$$10 = x \cdot 39.95$$

$$\frac{10}{39.95} = x$$

$$0.25 \approx x$$

$$25\% \approx x$$

The rate of discount is approximately 25%.

35. Let x = the regular price of the ring. The discount amount is $20\% \cdot x = 0.20x$. The sale price is $x - 0.2x = 0.8x$. The sale price is also $149.99, so

$$0.8x = 149.99$$

$$x = \frac{149.99}{0.8}$$

$$x = 187.4875$$

The regular price of the ring is $187.49.

37. The sale price is $399.97 - \$50 = \$349.97$. The discount is $50. $50 is what percent of $399.97?

$$50 = x \cdot 399.97$$

$$\frac{50}{399.97} = x$$

$$0.13 \approx x$$

$$13\% \approx x$$

The discount rate is approximately 13%.

39. The discount is $3.60. $3.60 is what percent of $15.48?

$$3.6 = x \cdot 15.48$$

$$\frac{3.6}{15.48} = x$$

$$0.23 \approx x$$

$$23\% \approx x$$

The discount rate is approximately 23%. The reduced price is $15.48 - \$3.60 = \$11.88$.

REVIEW

45. $-5(-5)(-2) = [-5(-5)](-2)$
$= (25)(-2)$
$= -50$

47. $2a^2(a+b) = 2(-2)^2[-2+(-1)]$
$= 2(4)(-2-1)$
$= 2(4)(-3)$
$= 8(-3)$
$= -24$

49. One dozen eggs is 12 eggs, so d dozen eggs is $12d$ eggs.

51. $|-5 - 8| = |-13|$
$= 13$

STUDY SET Estimation

1. 10% of 815 is 81.5, or about 82. To find 20%, double that. Thus, 20% of 815 is approximately $2 \cdot 82 = 164$.

3. 10% of 196.88 is 19.688 or about 20. To find 30%, triple that. Thus, the price of the VCR is discounted about $60.

5. 107,809 is about 108,000. Half of 108,000 is 54,000. The insurance company paid approximately $54,000.

7. 31% is about 30%. 10% of 68 is 6.8, which is about 7. To find 30%, triple that. Approximately $3 \cdot 7 = 21$ people actually attended the seminar.

9. 48% is about 50%. Half of 6,200 is 3,100. Approximately 3,100 volunteers helped with the election.

KEY CONCEPT Equivalent Expressions

1. $\dfrac{10}{24} = \dfrac{5}{12}$

The fraction was simplified.

3. $2x + 3x = 5x$
Like terms were combined.

5. $x^3 \cdot x^2 = x^{3+2}$
$\qquad = x^5$
When multiplying like bases, add the exponents.

7. $\dfrac{2}{3} \cdot \dfrac{3}{2} = 1$
The product of a number and its reciprocal is 1.

9. $4x^2 + 1 - 2x^2 = \left(4x^2 - 2x^2\right) + 1$
$\qquad = 2x^2 + 1$
Like terms were combined.

11. $\dfrac{6}{6} = 1$
A number divided by itself is 1.

13. $(-5)(-6) = 30$
The product of two negative numbers is positive.

15. $\dfrac{2x}{2} = x$
2 times x, divided by 2, is x.

17. $2 + 3 \cdot 5 = 2 + 15$
$\qquad = 17$
Multiplications are done before additions.

19. $2(x + 5) = 2 \cdot x + 2 \cdot 5$
$\qquad = 2x + 10$
The distributive property was used.

STUDY SET Section 7.4

VOCABULARY

1. In banking, the original amount of money borrowed or deposited is known as the <u>principal</u>.

3. The percent that is used to calculate the amount of interest to be paid is called the <u>interest</u> rate.

5. Interest computed only on the original principal is called <u>simple</u> interest.

CONCEPTS

7. a. 2 times per year

   b. 4 times per year

   c. 365 times per year

   d. 12 times per year

9. a. $7\% = 7.0\%$
$\qquad = .070$
$\qquad = 0.07$

   b. $9.8\% = .098$
$\qquad = 0.098$

   c. $6\dfrac{1}{4}\% = 6.25\%$
$\qquad = .0625$
$\qquad = 0.0625$

11. a. The diagram in Illustration 1 illustrates the concept of compound interest.

   b. The original principal was $1,000.

   c. The interest was found 4 times.

   d. The interest earned on the first compounding was $1,050 - $1,000 = $50.

   e. The money was invested for 4 quarters, which is 1 year.

NOTATION

13. *Prt* indicates multiplication of the numbers represented by *P*, *r*, and *t*.

APPLICATIONS

15. $P = 5,000; r = 6\% = 0.06; t = 1$
$I = Prt$
$I = 5,000 \cdot 0.06 \cdot 1$
$I = 300$
At the end of the first year, the account balance will be
$\$5,000 + \$300 = \$5,300$.

17. $P = 8,000; r = 9.2\% = 0.092;$
$t = 2$
$I = Prt$
$I = 8,000 \cdot 0.092 \cdot 2$
$I = 1,472$
$\$1,472$ in interest will be paid on the loan.

19. $P = 4,200; r = 18\% = 0.18;$
$t = \dfrac{30 \text{ days}}{365 \text{ days}} = \dfrac{6}{73}$
$I = Prt$
$I = 4,200 \cdot 0.18 \cdot \dfrac{6}{73}$
$I = \dfrac{4,200}{1} \cdot \dfrac{0.18}{1} \cdot \dfrac{6}{73}$
$I = \dfrac{4536}{73}$
$I \approx 62.14$
The business had to repay
$\$4,200 + \$62.14 = \$4,262.14$.

21. $P = 10,000; r = 7\dfrac{1}{4}\% = 7.25\%$
$= 0.0725;$
$t = 2$
$I = Prt$
$I = 10,000 \cdot 0.0725 \cdot 2$
$I = 1,450$

P	r	t	I
$\$10,000$	0.0725	2 yr	$\$1,450$

23.

Loan Application Worksheet
1. Amount of loan (principal) __$1,200.00__
2. Length of loan (time) __2 YEARS__
3. Annual percentage rate __8%__
4. Interest charged __$192__
5. Total amount to be repaid __$1,392__
6. Check method of repayment: ☐ 1 lump sum ☑ monthly payments
Borrower agrees to pay __24__ equal payments of __$58__ to repay loan.

25. $P = 18; r = 2.3\% = 0.023; t = 2$
$I = Prt$
$I = 18 \cdot 0.023 \cdot 2$
$I = 0.828$
The country must pay back
$\$18 + \$0.828 = \$18.828$ million.

27. $P = 600$; $r = 8\% = 0.08$; $t = 3$; $n = 1$

$$A = P\left(1 + \frac{r}{n}\right)^{nt}$$

$$A = 600\left(1 + \frac{0.08}{1}\right)^{1 \cdot 3}$$

$$A = 600(1 + 0.08)^3$$

$$A \approx 755.83$$

The account balance will be $755.83.

29. $P = 1{,}000$; $r = 6\% = 0.06$; $t = 4$; $n = 365$

$$A = P\left(1 + \frac{r}{n}\right)^{nt}$$

$$A = 1{,}000\left(1 + \frac{0.06}{365}\right)^{365 \cdot 4}$$

$$A = 1{,}000\left(1 + \frac{0.06}{365}\right)^{1{,}460}$$

$$A \approx 1271.22$$

The account will hold $1,271.22.

31. $P = 545$; $r = 4.6\% = 0.046$; $t = 1$; $n = 365$

$$A = P\left(1 + \frac{r}{n}\right)^{nt}$$

$$A = 545\left(1 + \frac{0.046}{365}\right)^{365 \cdot 1}$$

$$A = 545\left(1 + \frac{0.046}{365}\right)^{365}$$

$$A \approx 570.65$$

The account will contain $570.65.

33. $P = 500{,}000$; $r = 6\% = 0.06$; $t = 1$; $n = 365$

$$A = P\left(1 + \frac{r}{n}\right)^{nt}$$

$$A = 500{,}000\left(1 + \frac{0.06}{365}\right)^{365 \cdot 1}$$

$$A = 500{,}000\left(1 + \frac{0.06}{365}\right)^{365}$$

$$A \approx 530{,}915.66$$

The interest earned is the final account balance minus the original deposit, or
$530,915.66 − $500,000
= $30,915.66.

REVIEW

39. $$\sqrt{\frac{1}{4}} = \sqrt{\frac{1}{2} \cdot \frac{1}{2}}$$

$$= \frac{1}{2}$$

41. $y = 2x - 10$
$-3 \stackrel{?}{=} 2(2) - 10$
$-3 \stackrel{?}{=} 4 - 10$
$-3 \neq -6$
The point is not on the line.

43. $$\frac{2}{3}x = -2$$

$$3\left(\frac{2}{3}x\right) = 3(-2)$$

$$2x = -6$$

$$x = -3$$

45. $2x^2 - 3x + 5$ contains 3 terms.

CHAPTER 7 REVIEW

1. a. 39 of the 100 squares are shaded. This is 39%, 0.39, or $\dfrac{39}{100}$.

b. 100 of the boxes on the left side are shaded and 11 of the boxes on the right side are shaded. This is 111%, 1.11, or $\dfrac{111}{100} = 1\dfrac{11}{100}$.

3. a. $15\% = \dfrac{15}{100}$

$= \dfrac{3}{20}$

b. $120\% = \dfrac{120}{100}$

$= \dfrac{6}{5}$

c. $9\dfrac{1}{4}\% = \dfrac{9\frac{1}{4}}{100}$

$= 9\dfrac{1}{4} \div 100$

$= \dfrac{37}{4} \cdot \dfrac{1}{100}$

$= \dfrac{37}{400}$

3. d. $0.1\% = \dfrac{0.1}{100}$

$= 0.1 \div 100$

$= \dfrac{1}{10} \cdot \dfrac{1}{100}$

$= \dfrac{1}{1000}$

5. a. $0.83 = 83\%$

b. $0.625 = 62.5\%$

c. $0.051 = 5.1\%$

d. $6 = 600\%$

7. a.
$$\begin{array}{r} 0.33 \\ 3\overline{)1.00} \\ 9 \\ \hline 10 \\ 9 \\ \hline 1 \end{array}$$

$\dfrac{1}{3} = 0.\overline{3}$

$= 0.33\overline{3}$

$= 33.\overline{3}\%$

$= 33\dfrac{1}{3}\%$

7. b.

$$6\overline{)5.000} \quad 0.833$$
$$\underline{4\,8}$$
$$20$$
$$\underline{18}$$
$$20$$
$$\underline{18}$$
$$2$$

$$\frac{5}{6} = 0.8\overline{3}$$
$$= 0.83\overline{3}$$
$$= 83.\overline{3}\%$$
$$= 83\frac{1}{3}\%$$

9. The number of amendments adopted after the Bill of Rights is $27 - 10 = 17$. 17 is what percent of 27?

$$17 = x \cdot 27$$

$$\frac{17}{27} = x$$

$$0.63 \approx x$$

$$63\% \approx x$$

63% of the amendments were adopted after the Bill of Rights.

11. $x = 32\% \cdot 96$

13. 96% of 15 gallons is what number?

$$96\% \cdot 15 = x$$
$$0.96 \cdot 15 = x$$
$$14.4 = x$$

Thus there are 14.4 gallons of nitro in the fuel tank. Since the tank holds 15 gallons, there are $15 - 14.4 = 0.6$ gallons of methane in the tank. (This result can also be found by finding the number that is 4% of 15.)

15. 96 is what percent of 110?

$$96 = x \cdot 110$$

$$\frac{96}{110} = x$$

$$0.87 \approx x$$

$$87\% \approx x$$

Approximately 87% of the trailers were damaged.

17.

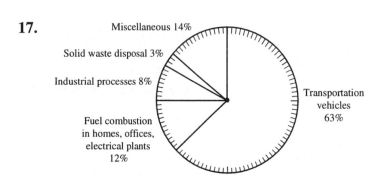

Miscellaneous 14%

Solid waste disposal 3%

Industrial processes 8%

Fuel combustion in homes, offices, electrical plants 12%

Transportation vehicles 63%

19. Sales tax: what number is 5.5% of $59.99?

$x = 5.5\% \cdot 59.99$

$x = 0.055 \cdot 59.99$

$x = 3.29945$

The sales tax is $3.30. The total is $59.99 + $3.30 = $63.29.

21. Together, the washing machine and dryer cost $369.97 + $299.97 = $669.94. What number is 6% of $669.94?

$x = 6\% \cdot 669.94$

$x = 0.06 \cdot 669.94$

$x = 40.1964$

The commission is $40.20.

23. The amount of the decrease is $18.8 - 17.0 = 1.8$. 1.8 is what percent of 18.8?

$1.8 = x \cdot 18.8$

$\dfrac{1.8}{18.8} = x$

$0.096 \approx x$

$9.6\% \approx x$

The percent of decrease is approximately 9.6%.

25. $P = 6,000$; $r = 8\% = 0.08$;

$t = 2$

$I = Prt$

$I = 6,000 \cdot 0.08 \cdot 2$

$I = 960$

$960 in interest is earned.

P	r	t	I
$6,000	8%	2 years	$960

27. First, find the total amount owed.

$P = 1,500;\ r = 7\dfrac{3}{4}\% = 7.75\% = 0.0775;$

$t = 1$

$I = Prt$

$I = 1,500 \cdot 0.0775 \cdot 1$

$I = 116.25$

The total amount the couple must repay is $1,500 + $116.25 = $1,616.25. Since the couple is making 12 equal monthly payments, each payment will be $\dfrac{1}{12}$ of the total amount.

$\dfrac{1}{12} \cdot 1,616.25 = \dfrac{1}{12} \cdot \dfrac{1,616.25}{1}$

$= \dfrac{1,616.25}{12}$

≈ 134.69

Each payment will be $134.69

29. $P = 5,000;$

$r = 6\dfrac{1}{2}\% = 6.5\% = 0.065;$

$t = 3;$

$n = 365$

$A = P\left(1 + \dfrac{r}{n}\right)^{nt}$

$A = 5,000\left(1 + \dfrac{0.065}{365}\right)^{365 \cdot 3}$

$A = 5,000\left(1 + \dfrac{0.065}{365}\right)^{1,095}$

$A \approx 6,076.45$

The account will hold $6,076.45.

CHAPTER 7 TEST

1. 100 of the boxes on the left side are shaded and 99 of the boxes on the right side are shaded. This is 199%, $\frac{199}{100} = 1\frac{99}{100}$, or 1.99.

3. a. $67\% = 0.67$

   b. $12.3\% = 0.123$

   c. $9\frac{3}{4}\% = 9.75\%$
      $= 0.0975$

5. a. $0.19 = 19\%$

   b. $3.47 = 347\%$

   c. $0.005 = 0.5\%$

7.
$$\begin{array}{r} 0.233 \\ 30\overline{)7.000} \\ \underline{6\ 0} \\ 1\ 00 \\ \underline{90} \\ 100 \\ \underline{90} \\ 10 \end{array}$$

   $\frac{7}{30} = 0.2\overline{3}$
   $= 0.23333$
   $= 23.33\overline{3}\%$
   $\approx 23.33\%$

9.
$$\begin{array}{r} 0.66 \\ 3\overline{)2.00} \\ \underline{1\ 8} \\ 20 \\ \underline{18} \\ 2 \end{array}$$

   $\frac{2}{3} = 0.\overline{6}$
   $= 0.66\overline{6}$
   $= 66.\overline{6}\%$
   $= 66\frac{2}{3}\%$

11. a. The length lost will be from the inseam. 3% of 34 inches is what number?
      $3\% \cdot 34 = x$
      $0.03 \cdot 34 = x$
      $1.02 = x$
      1.02 inches will be lost due to shrinkage.

    b. The resulting length will be
       34 inches − 1.02 inches
       = 32.98 inches.

13. What number is 15% of $25.40?
    $x = 15\% \cdot 25.40$
    $x = 0.15 \cdot 25.40$
    $x = 3.81$
    The tip should be $3.81.

15. 18 is 20% of what number?
    $18 = 20\% \cdot x$
    $18 = 0.20 \cdot x$
    $\frac{18}{0.20} = x$
    $90 = x$
    The swimmer normally completes 90 laps.

17. $x = 24\% \cdot 600$
$x = 0.24 \cdot 600$
$x = 144$

19. What number is 4% of $898?
$x = 4\% \cdot 898$
$x = 0.04 \cdot 898$
$x = 35.92$
Her commission is $35.92.

21. The sale price is $14.95 – $3
= $11.95. The discount is $3. $3
is what percent of $14.95?
$$3 = x \cdot 14.95$$

$$\frac{3}{14.95} = x$$

$$0.20 \approx x$$

$$20\% \approx x$$
The rate of discount is
approximately 20%.

23. $P = 3{,}000; r = 5\% = 0.05; t = 1$
$I = Prt$
$I = 3{,}000 \cdot 0.05 \cdot 1$
$I = 150$
The interest is $150.

STUDY SET Section 8.1

VOCABULARY

1. A ratio is a <u>comparison</u> of two numbers by their indicated <u>quotient</u>.

3. The ratios $\dfrac{2}{3}$ and $\dfrac{4}{6}$ are <u>equal</u> ratios.

CONCEPTS

5. Football statistics list the ratio of passes that the quarterback completes to the number of passes attempted. Answers will vary.

NOTATION

7. $\dfrac{2}{5}$

PRACTICE

9. $\dfrac{5}{7}$

11. $\dfrac{17}{34} = \dfrac{17}{2 \cdot 17}$

$= \dfrac{1}{2}$

13. $\dfrac{22}{33} = \dfrac{2 \cdot 11}{3 \cdot 11}$

$= \dfrac{2}{3}$

15. $\dfrac{7}{24.5} = \dfrac{7 \cdot 10}{24.5 \cdot 10}$

$= \dfrac{70}{245}$

$= \dfrac{2 \cdot 35}{7 \cdot 35}$

$= \dfrac{2}{7}$

17. $\dfrac{4 \text{ ounces}}{12 \text{ ounces}} = \dfrac{4}{4 \cdot 3}$

$= \dfrac{1}{3}$

19. $\dfrac{12 \text{ minutes}}{1 \text{ hour}} = \dfrac{12 \text{ minutes}}{60 \text{ minutes}}$

$= \dfrac{12}{12 \cdot 5}$

$= \dfrac{1}{5}$

21. $\dfrac{3 \text{ days}}{1 \text{ week}} = \dfrac{3 \text{ days}}{7 \text{ days}}$

$= \dfrac{3}{7}$

23. $\dfrac{18 \text{ months}}{2 \text{ years}} = \dfrac{18 \text{ months}}{2 \cdot 12 \text{ months}}$

$= \dfrac{18 \text{ months}}{24 \text{ months}}$

$= \dfrac{6 \cdot 3}{6 \cdot 4}$

$= \dfrac{3}{4}$

25. The total amount of the budget is $800 + $600 + $180 + $100 + $120 = $1,800.

27. $\dfrac{\text{food}}{\text{total}} = \dfrac{600}{1,800}$

$\phantom{\dfrac{\text{food}}{\text{total}}} = \dfrac{600}{3 \cdot 600}$

$\phantom{\dfrac{\text{food}}{\text{total}}} = \dfrac{1}{3}$

29. The total amount of the deductions is $995 + $1,245 + $1,680 + $4,580 + $225 = $8,725.

31. $\dfrac{\text{charitable contributions}}{\text{total deductions}} = \dfrac{1,680}{8,725}$

$\phantom{\dfrac{\text{charitable contributions}}{\text{total deductions}}} = \dfrac{5 \cdot 336}{5 \cdot 1,745}$

$\phantom{\dfrac{\text{charitable contributions}}{\text{total deductions}}} = \dfrac{336}{1,745}$

APPLICATIONS

33. $\dfrac{\text{faculty}}{\text{students}} = \dfrac{125}{2,000}$

$\phantom{\dfrac{\text{faculty}}{\text{students}}} = \dfrac{125}{125 \cdot 16}$

$\phantom{\dfrac{\text{faculty}}{\text{students}}} = \dfrac{1}{16}$

35. $\dfrac{\$21.59}{17 \text{ gal}} = 1.27 \dfrac{\text{dollars}}{\text{gallon}}$

$\phantom{\dfrac{\$21.59}{17 \text{ gal}}} = \$1.27 \text{ per gallon}$

37. $\dfrac{84 \text{ cents}}{12 \text{ ounces}} = 7\text{¢ per oz}$

39. The unit cost for the 6-ounce can is $\dfrac{89 \text{ cents}}{6 \text{ ounces}} = 14.8\overline{3}\text{¢ per oz.}$ The unit cost for the 8-ounce can is $\dfrac{119 \text{ cents}}{8 \text{ ounces}} = 14.875\text{¢ per oz.}$ The 6-oz can has a lower unit cost so it is a better buy.

41. The first student read $\dfrac{54 \text{ pages}}{40 \text{ minutes}} = 1.35 \text{ pages per minute.}$ The second student read $\dfrac{80 \text{ pages}}{62 \text{ minutes}} \approx 1.29 \text{ pages per minute.}$ The first student read faster.

43. $\dfrac{11,880 \text{ gallons}}{27 \text{ minutes}} = 440 \text{ gallons per minute}$

45. The amount of sales tax was $36.75 - $35 = $1.75.
The tax rate is the ratio of the amount of sales tax to the original price of the sweater.

$$\frac{\$1.75}{\$35} = 0.05$$

The sales tax rate is 5%.

47. The car has traveled 35,071 − 34,746 = 325 miles.

The average rate of speed is $\dfrac{325 \text{ miles}}{5 \text{ hours}} = 65$ mph.

49. The gas mileage for the first car is $\dfrac{1,235 \text{ miles}}{51.3 \text{ gallons}} \approx 24.1$ miles per gallon.

The gas mileage for the second car is $\dfrac{1,456 \text{ miles}}{55.78 \text{ gallons}} \approx 26.1$ miles per gallon.

The second car got better gas mileage.

REVIEW

53. $3.05 + 17.17 + 25.317 = 20.22 + 25.317$
$$= 45.537$$

55. $13.2 + 25.07 \cdot 7.16 = 13.2 + 179.5012$
$$= 192.7012$$

57. $5 - 3\dfrac{1}{4} = \dfrac{5}{1} - \dfrac{13}{4}$

$$= \dfrac{5 \cdot 4}{1 \cdot 4} - \dfrac{13}{4}$$

$$= \dfrac{20}{4} - \dfrac{13}{4}$$

$$= \dfrac{20 - 13}{4}$$

$$= \dfrac{7}{4}$$

$$= 1\dfrac{3}{4}$$

59. $3x - 2 = 19$
$3x = 21$
$x = 7$

KEY CONCEPT Problem Solving

You are asked to find <u>the cost per day of the cruise</u>.
There is <u>1</u> unknown.
Let $x =$ <u>the cost per day of the cruise</u>.

The number of days of the cruise	times	<u>the cost per day</u>	is	the total cost of the cruise.

$$5x = \underline{995.95}$$

$$\frac{5x}{5} = \frac{995.95}{5} \quad \text{Divide both sides by } \underline{5}.$$

$$x = \underline{199.19}$$

<u>The cruise costs $199.19 per day.</u>
The cost per day is $\underline{\$199.19}$, and the cruise lasts for $\underline{5}$ days.

STUDY SET Section 8.2

VOCABULARY

1. A <u>proportion</u> is a statement that two <u>ratios</u> are equal.

3. The second and third terms of a proportion are called the <u>means</u> of the proportion.

CONCEPTS

5. The equation $\dfrac{a}{b} = \dfrac{c}{d}$ will be a proportion if the product <u>ad</u> is equal to the product <u>bc</u>.

NOTATION

7. $\quad 12 \cdot 24 = \underline{18x}$

$$\underline{288} = 18x$$

$$\frac{288}{18} = \frac{18x}{18}$$

$$16 = x$$

PRACTICE

9. The product of the extremes is $9 \cdot 70 = 630$.
The product of the means is $7 \cdot 81 = 567$.
Since $630 \neq 567$, the statement is not a proportion.

11. The product of the extremes is $7 \cdot 6 = 42$.
The product of the means is $3 \cdot 14 = 42$.
Since $42 = 42$, the statement is a proportion.

13. The product of the extremes is $9 \cdot 80 = 720$.
The product of the means is $19 \cdot 38 = 722$.
Since $720 \neq 722$, the statement is not a proportion.

15. The product of the extremes is
$10.4 \cdot 14.4 = 149.76$.
The product of the means is
$3.6 \cdot 41.6 = 149.76$.
Since $149.76 = 149.76$, the
statement is a proportion.

17.
$$\frac{2}{3} = \frac{x}{6}$$
$$2 \cdot 6 = 3 \cdot x$$
$$12 = 3x$$
$$4 = x$$

19.
$$\frac{5}{10} = \frac{3}{c}$$
$$5 \cdot c = 10 \cdot 3$$
$$5c = 30$$
$$c = 6$$

21.
$$\frac{6}{x} = \frac{8}{4}$$
$$6 \cdot 4 = x \cdot 8$$
$$24 = 8x$$
$$3 = x$$

23.
$$\frac{x}{3} = \frac{9}{3}$$
$$x \cdot 3 = 3 \cdot 9$$
$$3x = 27$$
$$x = 9$$

25.
$$\frac{x+1}{5} = \frac{3}{15}$$
$$(x+1)15 = 5 \cdot 3$$
$$15(x+1) = 15$$
$$15x + 15 = 15$$
$$15x = 0$$
$$x = 0$$

27.
$$\frac{x+3}{12} = \frac{-7}{6}$$
$$(x+3)6 = 12(-7)$$
$$6(x+3) = -84$$
$$6x + 18 = -84$$
$$6x = -102$$
$$x = -\frac{102}{6}$$
$$x = -17$$

29.
$$\frac{4-x}{13} = \frac{11}{26}$$
$$(4-x)26 = 13 \cdot 11$$
$$26(4-x) = 143$$
$$104 - 26x = 143$$
$$-26x = 39$$
$$x = \frac{39}{-26}$$
$$x = -\frac{3}{2}$$

31.
$$\frac{2x+1}{18} = \frac{14}{3}$$
$$(2x+1)3 = 18 \cdot 14$$
$$3(2x+1) = 252$$
$$6x + 3 = 252$$
$$6x = 249$$
$$x = \frac{249}{6}$$
$$x = \frac{83}{2}$$

APPLICATIONS

33. 3 pints is to $1 as 51 pints is to c.

$$\frac{3}{1} = \frac{51}{c}$$

$$3c = 1 \cdot 51$$

$$3c = 51$$

$$c = 17$$

51 pints cost $17.

35. 3 packets is to 50¢ as 39 packets is to c¢.

$$\frac{3}{50} = \frac{39}{c}$$

$$3c = 50 \cdot 39$$

$$3c = 1950$$

$$c = 650$$

39 packets cost 650¢ or $6.50.

37. The ratio of costs to revenue in 1995 is $\frac{4}{10} = \frac{2}{5}$. The ratio of costs to revenue in 1996 is $\frac{10}{25} = \frac{2}{5}$. The ratios of costs to revenue is the same for both years.

39. 3 drops of essence is to 7 drops of alcohol as d drops of essence is to 56 drops of alcohol.

$$\frac{3}{7} = \frac{d}{56}$$

$$3 \cdot 56 = 7d$$

$$168 = 7d$$

$$24 = d$$

24 drops of essence should be used.

41. $1\frac{1}{4} = 1.25$ cups of flour is to

$3\frac{1}{2} = 3.5$ dozen cookies as c cups of flour is to 12 dozen cookies.

$$\frac{1.25}{3.5} = \frac{c}{12}$$

$$1.25 \cdot 12 = 3.5c$$

$$15 = 3.5c$$

$$\frac{15}{3.5} = c$$

$$4.29 \approx c$$

About $4\frac{1}{4}$ cups of flour should be used.

43. 95% is 95 per 100. Thus, 5 parts per 100 are expected to be defective. 5 defective is to 100 parts as p defective is to 940 parts.

$$\frac{5}{100} = \frac{p}{940}$$

$$5 \cdot 940 = 100p$$

$$4,700 = 100p$$

$$47 = p$$

In a run of 940 pieces, 47 defective parts are expected.

45. 42 miles is to 1 gallon as 315 miles is to g gallons.

$$\frac{42}{1} = \frac{315}{g}$$

$$42g = 315$$

$$g = \frac{315}{42}$$

$$g = 7.5$$

7.5 gallons of gas are needed.

47. If Bill missed 10 hours of work, he worked 30 hours. $412 is to 40 hours as $d is to 30 hours.

$$\frac{412}{40} = \frac{d}{30}$$

$$412 \cdot 30 = 40d$$

$$12,360 = 40d$$

$$309 = d$$

He got paid $309.

49. 169 feet is to 1 foot as c inches is to 3.5 inches.

$$\frac{169}{1} = \frac{c}{3.5}$$

$$169 \cdot 3.5 = c$$

$$591.5 = c$$

A real caboose is $591\frac{1}{2}$ inches or

49 ft $3\frac{1}{2}$ in. long.

51. For Glenwood High, 3 teachers is to 50 students as t teachers is to 2,700 students.

$$\frac{3}{50} = \frac{t}{2,700}$$

$$3 \cdot 2,700 = 50t$$

$$8,100 = 50t$$

$$162 = t$$

For Goddard Junior High, 3 teachers is to 50 students as t teachers is to 1,900 students.

$$\frac{3}{50} = \frac{t}{1,900}$$

$$3 \cdot 1,900 = 50t$$

$$5,700 = 50t$$

$$114 = t$$

For Sellers Elementary, 3 teachers is to 50 students as t teachers is to 850 students.

$$\frac{3}{50} = \frac{t}{850}$$

$$3 \cdot 850 = 50t$$

$$2,550 = 50t$$

$$51 = t$$

	Glenwood High	Goddard Junior High	Sellers Elementary
Enrollment	2,700	1,900	850
Teachers	162	114	51

53. 6 gallons is 6 • 128 = 768 ounces. We can use a proportion to find out how many ounces of oil should be used for 6 gallons of gas. 50 is to 1 as 768 ounces is to x ounces.

$$\frac{50}{1} = \frac{768}{x}$$

$$50x = 768$$

$$x = \frac{768}{50}$$

$$x = 15.36$$

The instructions are not exactly correct, but they are close.

REVIEW

57. $\dfrac{9}{10} = 9 \div 10$

$$= 0.9$$

$$= 90\%$$

59. $33\dfrac{1}{3}\% = \dfrac{33\frac{1}{3}}{100}$

$$= 33\frac{1}{3} \div 100$$

$$= \frac{100}{3} \cdot \frac{1}{100}$$

$$= \frac{1\cancel{00}}{3 \cdot 1\cancel{00}}$$

$$= \frac{1}{3}$$

61. What number is 30% of 1,600?

$x = 30\% \cdot 1,600$

$x = 0.30 \cdot 1,600$

$x = 480$

63. What number is 25% of $98?

$x = 25\% \cdot 98$

$x = 0.25 \cdot 98$

$x = 24.5$

This is the amount of the discount, so Maria paid $98 – $24.5 = $73.50.

65. The discount amount on the shoes was $30\% \cdot 59 = 0.30 \cdot 59$
$= 17.7$
He paid $59 – $17.7 = $41.30 for the shoes. The discount amount on the boots was $40\% \cdot 79 = 0.40 \cdot 79$
$= 31.6$
He paid $79 – $31.6 = $47.40 for the boots.
Ricardo paid a total of $41.30 + $47.40 = $88.70 for the footwear.

STUDY SET Section 8.3

VOCABULARY

1. Units of inches, feet, and miles are examples of <u>American</u> units of length.

3. The value of any unit conversion factor is <u>1</u>.

5. Some examples of American units of capacity are <u>cups</u>, <u>pints</u>, <u>quarts</u>, and <u>gallons</u>.

CONCEPTS

7. 12 in. = <u>1</u> ft

9. 1 mi = <u>5,280</u> ft

11. <u>16</u> ounces = 1 pound

13. 1 cup = <u>8</u> fluid ounces

15. 2 pints = <u>1</u> quart

17. 1 day = <u>24</u> hours

NOTATION

19. $12 \text{ yd} = 12 \text{ yd} \cdot \dfrac{36 \text{ in.}}{1 \text{ yd}}$

$\qquad = 12 \cdot \underline{36} \text{ in.}$

$\qquad = 432 \text{ in.}$

21. $12 \text{ pt} = 12 \text{ pt} \cdot \dfrac{1 \text{ qt}}{\underline{2} \text{ pt}} \cdot \dfrac{1 \text{ gal}}{\underline{4} \text{ qt}}$

$\qquad = \underline{12} \cdot \dfrac{1}{2} \cdot \dfrac{1}{4} \text{ gal}$

$\qquad = 1.5 \text{ gal}$

PRACTICE

23. The width of a dollar bill is about $2\dfrac{5}{8}$ in.

25. The diameter of a quarter is about 1 in.

27. The length of a sheet of typing paper is 11 in.

29. The height of a Coca-Cola can is about $4\dfrac{3}{4}$ in.

31. $4 \text{ ft} = 4 \text{ ft} \cdot \dfrac{12 \text{ in.}}{1 \text{ ft}}$

$\qquad = 4 \cdot 12 \text{ in.}$

$\qquad = 48 \text{ in.}$

33. $3\dfrac{1}{2} \text{ ft} = 3\dfrac{1}{2} \text{ ft} \cdot \dfrac{12 \text{ in.}}{1 \text{ ft}}$

$\qquad = 3\dfrac{1}{2} \cdot 12 \text{ in.}$

$\qquad = \dfrac{7}{2} \cdot 12 \text{ in.}$

$\qquad = 42 \text{ in.}$

35. $24 \text{ in.} = 24 \text{ in.} \cdot \dfrac{1 \text{ ft}}{12 \text{ in.}}$

$\qquad = 24 \cdot \dfrac{1}{12} \text{ ft}$

$\qquad = 2 \text{ ft}$

37. $8 \text{ yd} = 8 \text{ yd} \cdot \dfrac{3 \text{ ft}}{1 \text{ yd}} \cdot \dfrac{12 \text{ in.}}{1 \text{ ft}}$

$\qquad = 8 \cdot 3 \cdot 12 \text{ in.}$

$\qquad = 288 \text{ in.}$

39. $90 \text{ in.} = 90 \text{ in.} \cdot \dfrac{1 \text{ ft}}{12 \text{ in.}} \cdot \dfrac{1 \text{ yd}}{3 \text{ ft}}$

$\qquad = 90 \cdot \dfrac{1}{12} \cdot \dfrac{1}{3} \text{ yd}$

$\qquad = 2.5 \text{ yd}$

41. $56 \text{ in.} = 56 \text{ in.} \cdot \dfrac{1 \text{ ft}}{12 \text{ in.}}$

$\qquad = 56 \cdot \dfrac{1}{12} \text{ ft}$

$\qquad = \dfrac{14}{3} \text{ ft}$

$\qquad = 4\dfrac{2}{3} \text{ ft}$

43. $5 \text{ yd} = 5 \text{ yd} \cdot \dfrac{3 \text{ ft}}{1 \text{ yd}}$

$\qquad = 5 \cdot 3 \text{ ft}$

$\qquad = 15 \text{ ft}$

45. $7 \text{ ft} = 7 \text{ ft} \cdot \dfrac{1 \text{ yd}}{3 \text{ ft}}$

$= 7 \cdot \dfrac{1}{3} \text{ yd}$

$= \dfrac{7}{3} \text{ yd}$

$= 2\dfrac{1}{3} \text{ yd}$

47. $15{,}840 \text{ ft} = 15{,}840 \text{ ft} \cdot \dfrac{1 \text{ mi}}{5{,}280 \text{ ft}}$

$= 15{,}840 \cdot \dfrac{1}{5{,}280} \text{ mi}$

$= 3 \text{ mi}$

49. $\dfrac{1}{2} \text{ mi} = \dfrac{1}{2} \text{ mi} \cdot \dfrac{5{,}280 \text{ ft}}{1 \text{ mi}}$

$= \dfrac{1}{2} \cdot 5{,}280 \text{ ft}$

$= 2{,}640 \text{ ft}$

51. $80 \text{ oz} = 80 \text{ oz} \cdot \dfrac{1 \text{ lb}}{16 \text{ oz}}$

$= 80 \cdot \dfrac{1}{16} \text{ lb}$

$= 5 \text{ lb}$

53. $7{,}000 \text{ lb} = 7{,}000 \text{ lb} \cdot \dfrac{1 \text{ ton}}{2{,}000 \text{ lb}}$

$= 7{,}000 \cdot \dfrac{1}{2{,}000} \text{ tons}$

$= 3.5 \text{ tons}$

55. $12.4 \text{ tons} = 12.4 \text{ tons} \cdot \dfrac{2{,}000 \text{ lb}}{1 \text{ ton}}$

$= 12.4 \cdot 2{,}000 \text{ lb}$

$= 24{,}800 \text{ lb}$

57. $3 \text{ qt} = 3 \text{ qt} \cdot \dfrac{2 \text{ pt}}{1 \text{ qt}}$

$= 3 \cdot 2 \text{ pt}$

$= 6 \text{ pt}$

59. $16 \text{ pt} = 16 \text{ pt} \cdot \dfrac{1 \text{ qt}}{2 \text{ pt}} \cdot \dfrac{1 \text{ gal}}{4 \text{ qt}}$

$= 16 \cdot \dfrac{1}{2} \cdot \dfrac{1}{4} \text{ gal}$

$= 2 \text{ gal}$

61. $32 \text{ fl oz} = 32 \text{ fl oz} \cdot \dfrac{1 \text{ c}}{8 \text{ fl oz}} \cdot \dfrac{1 \text{ pt}}{2 \text{ c}}$

$= 32 \cdot \dfrac{1}{8} \cdot \dfrac{1}{2} \text{ pt}$

$= 2 \text{ pt}$

63. $240 \text{ min} = 240 \text{ min} \cdot \dfrac{1 \text{ hr}}{60 \text{ min}}$

$= 240 \cdot \dfrac{1}{60} \text{ hr}$

$= \dfrac{240}{1} \cdot \dfrac{1}{60} \text{ hr}$

$= \dfrac{240}{60} \text{ hr}$

$= 4 \text{ hr}$

65. $7{,}200 \text{ min} = 7{,}200 \text{ min} \cdot \dfrac{1 \text{ hr}}{60 \text{ min}} \cdot \dfrac{1 \text{ day}}{24 \text{ hr}}$

$= 7{,}200 \cdot \dfrac{1}{60} \cdot \dfrac{1}{24} \text{ days}$

$= \dfrac{7{,}200}{1} \cdot \dfrac{1}{60} \cdot \dfrac{1}{24} \text{ days}$

$= \dfrac{7{,}200}{1{,}440} \text{ days}$

$= 5 \text{ days}$

APPLICATIONS

67. Convert 450 feet to yards.

$$450 \text{ ft} = 450 \text{ ft} \cdot \frac{1 \text{ yd}}{3 \text{ ft}}$$

$$= 450 \cdot \frac{1}{3} \text{ yd}$$

$$= 150 \text{ yd}$$

The pyramid is about 150 yd high.

69. Convert 726 feet to miles.

$$726 \text{ ft} = 726 \text{ ft} \cdot \frac{1 \text{ mi}}{5,280 \text{ ft}}$$

$$= 726 \cdot \frac{1}{5,280} \text{ mi}$$

$$= \frac{726}{5,280} \text{ mi}$$

$$= 0.1375 \text{ mi}$$

Hoover Dam is 0.1375 mi high.

71. Convert 8 pounds to ounces.

$$8 \text{ lb} = 8 \text{ lb} \cdot \frac{16 \text{ oz}}{1 \text{ lb}}$$

$$= 8 \cdot 16 \text{ oz}$$

$$= 128 \text{ oz}$$

One gallon of water weighs about 128 oz.

73. Convert 5,000 pounds to tons.

$$5,000 \text{ lb} = 5,000 \text{ lb} \cdot \frac{1 \text{ ton}}{2,000 \text{ lb}}$$

$$= 5,000 \cdot \frac{1}{2,000} \text{ tons}$$

$$= \frac{5,000}{2,000} \text{ tons}$$

$$= 2.5 \text{ tons}$$

The average block weighed 2.5 tons.

75. Convert 17 gallons to quarts.

$$17 \text{ gal} = 17 \text{ gal} \cdot \frac{4 \text{ qt}}{1 \text{ gal}}$$

$$= 17 \cdot 4 \text{ qt}$$

$$= 68 \text{ qt}$$

He will need to buy 68 cans of paint.

77. Since there are 575 students attending the school, 575 pints of milk are used each day.
Convert 575 pints to gallons.

$$575 \text{ pt} = 575 \text{ pt} \cdot \frac{1 \text{ qt}}{2 \text{ pt}} \cdot \frac{1 \text{ gal}}{4 \text{ qt}}$$

$$= 575 \cdot \frac{1}{2} \cdot \frac{1}{4} \text{ gal}$$

$$= \frac{575}{1} \cdot \frac{1}{2} \cdot \frac{1}{4} \text{ gal}$$

$$= \frac{575}{8} \text{ gal}$$

$$= 71.875 \text{ gal}$$

The school uses 71.875 gallons of milk each day.

79. Convert 155 minutes to hours.

$$155 \text{ min} = 155 \text{ min} \cdot \frac{1 \text{ hr}}{60 \text{ min}}$$

$$= 155 \cdot \frac{1}{60} \text{ hr}$$

$$= \frac{155}{60} \text{ hr}$$

$$\approx 2.6 \text{ hr}$$

He walks about 2.6 hours.

REVIEW

83. 3,700

85. 3,673.26

87. 0.101

89. 0.1

STUDY SET Section 8.4

VOCABULARY

1. *Deka* means <u>tens</u>.

3. *Kilo* means <u>thousands</u>.

5. *Centi* means <u>hundredths</u>.

7. The amount of material in an object is called its <u>mass</u>.

CONCEPTS

9. 1 dekameter = <u>10</u> meters

11. 1 centimeter = $\dfrac{1}{\underline{100}}$ meter

13. 1 millimeter = $\dfrac{1}{\underline{1,000}}$ meter

15. 1 gram = <u>1,000</u> milligrams

17. 1 kilogram = <u>1,000</u> grams

19. 1 liter = <u>1,000</u> cubic centimeters

21. 1 centiliter = $\dfrac{1}{\underline{100}}$ liter

23. 100 liters = <u>1</u> hectoliter

NOTATION

25. $20 \text{ cm} = 20 \text{ cm} \cdot \dfrac{1 \text{ m}}{100 \text{ cm}}$

$= \dfrac{20}{100} \text{ m}$

$= 0.2 \text{ m}$

27. $2 \text{ km} = 2 \text{ km} \cdot \dfrac{1,000 \text{ m}}{1 \text{ km}} \cdot \dfrac{10 \text{ dm}}{1 \text{ m}}$

$= 2 \cdot 1,000 \cdot 10 \text{ dm}$

$= 20,000 \text{ dm}$

PRACTICE

29. The length of a dollar bill is about 156 mm.

31. The diameter of a nickel is about 21 mm.

33. The length of a sheet of typing paper is about 28 cm.

35. The length of one octave on a piano keyboard is about 16 cm.

37. $3 \text{ m} = 3 \text{ m} \cdot \dfrac{100 \text{ cm}}{1 \text{ m}}$

$= 3 \cdot 100 \text{ cm}$

$= 300 \text{ cm}$

39. $5.7 \text{ m} = 5.7 \text{ m} \cdot \dfrac{100 \text{ cm}}{1 \text{ m}}$

$= 5.7 \cdot 100 \text{ cm}$

$= 570 \text{ cm}$

41. $0.31 \text{ dm} = 0.31 \text{ dm} \cdot \dfrac{1 \text{ m}}{10 \text{ dm}} \cdot \dfrac{100 \text{ cm}}{1 \text{ m}}$

$\qquad = \dfrac{0.31 \cdot 100}{10} \text{ cm}$

$\qquad = 3.1 \text{ cm}$

43. $76.8 \text{ hm} = 76.8 \text{ hm} \cdot \dfrac{100 \text{ m}}{1 \text{ hm}} \cdot \dfrac{1{,}000 \text{ mm}}{1 \text{ m}}$

$\qquad = 76.8 \cdot 100 \cdot 1{,}000 \text{ mm}$

$\qquad = 7{,}680{,}000 \text{ mm}$

45. $4.72 \text{ cm} = 4.72 \text{ cm} \cdot \dfrac{1 \text{ m}}{100 \text{ cm}} \cdot \dfrac{10 \text{ dm}}{1 \text{ m}}$

$\qquad = \dfrac{4.72 \cdot 10}{100} \text{ dm}$

$\qquad = 0.472 \text{ dm}$

47. $453.2 \text{ cm} = 453.2 \text{ cm} \cdot \dfrac{1 \text{ m}}{100 \text{ cm}}$

$\qquad = \dfrac{453.2}{100} \text{ m}$

$\qquad = 4.532 \text{ m}$

49. $0.325 \text{ dm} = 0.325 \text{ dm} \cdot \dfrac{1 \text{ m}}{10 \text{ dm}}$

$\qquad = \dfrac{0.325}{10} \text{ m}$

$\qquad = 0.0325 \text{ m}$

51. $3.75 \text{ cm} = 3.75 \text{ cm} \cdot \dfrac{1 \text{ m}}{100 \text{ cm}} \cdot \dfrac{1{,}000 \text{ mm}}{1 \text{ m}}$

$\qquad = \dfrac{3.75 \cdot 1{,}000}{100} \text{ mm}$

$\qquad = 37.5 \text{ mm}$

53. $0.125 \text{ m} = 0.125 \text{ m} \cdot \dfrac{1{,}000 \text{ mm}}{1 \text{ m}}$

$\qquad = 0.125 \cdot 1{,}000 \text{ mm}$

$\qquad = 125 \text{ mm}$

55. $675 \text{ dam} = 675 \text{ dam} \cdot \dfrac{10 \text{ m}}{1 \text{ dam}} \cdot \dfrac{100 \text{ cm}}{1 \text{ m}}$

$\qquad\qquad = 675 \cdot 10 \cdot 100 \text{ cm}$

$\qquad\qquad = 675,000 \text{ cm}$

57. $638.3 \text{ m} = 638.3 \text{ m} \cdot \dfrac{1 \text{ hm}}{100 \text{ m}}$

$\qquad\qquad = \dfrac{638.3}{100} \text{ hm}$

$\qquad\qquad = 6.383 \text{ hm}$

59. $6.3 \text{ mm} = 6.3 \text{ mm} \cdot \dfrac{1 \text{ m}}{1,000 \text{ mm}} \cdot \dfrac{100 \text{ cm}}{1 \text{ m}}$

$\qquad\qquad = \dfrac{6.3 \cdot 100}{1,000} \text{ cm}$

$\qquad\qquad = 0.63 \text{ cm}$

61. $695 \text{ dm} = 695 \text{ dm} \cdot \dfrac{1 \text{ m}}{10 \text{ dm}}$

$\qquad\qquad = \dfrac{695}{10} \text{ m}$

$\qquad\qquad = 69.5 \text{ m}$

63. $5,689 \text{ m} = 5,689 \text{ m} \cdot \dfrac{1 \text{ km}}{1,000 \text{ m}}$

$\qquad\qquad = \dfrac{5,689}{1,000} \text{ km}$

$\qquad\qquad = 5.689 \text{ km}$

65. $576.2 \text{ mm} = 576.2 \text{ mm} \cdot \dfrac{1 \text{ m}}{1,000 \text{ mm}} \cdot \dfrac{10 \text{ dm}}{1 \text{ m}}$

$\qquad\qquad = \dfrac{576.2 \cdot 10}{1,000} \text{ dm}$

$\qquad\qquad = 5.762 \text{ dm}$

67. $6.45 \text{ dm} = 6.45 \text{ dm} \cdot \dfrac{1 \text{ m}}{10 \text{ dm}} \cdot \dfrac{1 \text{ km}}{1{,}000 \text{ m}}$

$\quad\quad = \dfrac{6.45}{10 \cdot 1{,}000} \text{ km}$

$\quad\quad = 0.000645 \text{ km}$

69. $658.23 \text{ m} = 658.23 \text{ m} \cdot \dfrac{1 \text{ km}}{1{,}000 \text{ m}}$

$\quad\quad = \dfrac{658.23}{1{,}000} \text{ km}$

$\quad\quad = 0.65823 \text{ km}$

71. $3 \text{ g} = 3 \text{ g} \cdot \dfrac{1{,}000 \text{ mg}}{1 \text{ g}}$

$\quad\quad = 3 \cdot 1{,}000 \text{ mg}$

$\quad\quad = 3{,}000 \text{ mg}$

73. $2 \text{ kg} = 2 \text{ kg} \cdot \dfrac{1{,}000 \text{ g}}{1 \text{ kg}}$

$\quad\quad = 2 \cdot 1{,}000 \text{ g}$

$\quad\quad = 2{,}000 \text{ g}$

75. $1 \text{ t} = 1 \text{ t} \cdot \dfrac{1{,}000 \text{ kg}}{1 \text{ t}} \cdot \dfrac{1{,}000 \text{ g}}{1 \text{ kg}}$

$\quad\quad = 1 \cdot 1{,}000 \cdot 1{,}000 \text{ g}$

$\quad\quad = 1{,}000{,}000 \text{ g}$

77. $500 \text{ mg} = 500 \text{ mg} \cdot \dfrac{1 \text{ g}}{1{,}000 \text{ mg}}$

$\quad\quad = \dfrac{500}{1{,}000} \text{ g}$

$\quad\quad = 0.5 \text{ g}$

79. $3 \text{ kL} = 3 \text{ kL} \cdot \dfrac{1{,}000 \text{ L}}{1 \text{ kL}}$

$\quad\quad = 3 \cdot 1{,}000 \text{ L}$

$\quad\quad = 3{,}000 \text{ L}$

81. $500 \text{ cL} = 500 \text{ cL} \cdot \dfrac{1 \text{ L}}{100 \text{ cL}} \cdot \dfrac{1{,}000 \text{ mL}}{1 \text{ L}}$

$\quad\quad = \dfrac{500 \cdot 1{,}000}{100} \text{ mL}$

$\quad\quad = 5{,}000 \text{ mL}$

83. $2 \text{ hL} = 2 \text{ hL} \cdot \dfrac{100 \text{ L}}{1 \text{ hL}} \cdot \dfrac{100 \text{ cL}}{1 \text{ L}}$

$\quad\quad = 2 \cdot 100 \cdot 100 \text{ cL}$

$\quad\quad = 20{,}000 \text{ cL}$

APPLICATIONS

85. Convert 50 meters to decimeters.

$\quad 50 \text{ m} = 50 \text{ m} \cdot \dfrac{10 \text{ dm}}{1 \text{ m}}$

$\quad\quad = 50 \cdot 10 \text{ dm}$

$\quad\quad = 500 \text{ dm}$

The swimming pool is 500 dm long.

87. Convert 4 kilograms to centigrams.

$\quad 4 \text{ kg} = 4 \text{ kg} \cdot \dfrac{1{,}000 \text{ g}}{1 \text{ kg}} \cdot \dfrac{100 \text{ cg}}{1 \text{ g}}$

$\quad\quad = 4 \cdot 1{,}000 \cdot 100 \text{ cg}$

$\quad\quad = 400{,}000 \text{ cg}$

The baby weighs 400,000 centigrams.

89. Two 1-liter bottles hold 2 liters. Convert 2 liters to deciliters.

$\quad 2 \text{ L} = 2 \text{ L} \cdot \dfrac{10 \text{ dL}}{1 \text{ L}}$

$\quad\quad = 2 \cdot 10 \text{ dL}$

$\quad\quad = 20 \text{ dL}$

The bottles hold 20 deciliters of root beer.

91. 1 kilogram is 1,000 grams. We need to find the number of 284-gram bottles it takes to get 1,000 grams of olives.

$\quad \dfrac{1{,}000 \text{ g}}{284 \text{ g}} \approx 3.5$

Thus, it takes 4 bottles of olives to get at least 1 kilogram of olives.

93. The 60 tablets contain 60 • 50 mg = 3,000 mg of active ingredient. Convert 3,000 milligrams to grams.

$\quad 3{,}000 \text{ mg} = 3{,}000 \text{ mg} \cdot \dfrac{1 \text{ g}}{1{,}000 \text{ mg}}$

$\quad\quad = \dfrac{3{,}000}{1{,}000} \text{ g}$

$\quad\quad = 3 \text{ g}$

The bottle contains 3 g of active ingredient.

REVIEW

97. 7% of \$342.72 is what number?

$\quad 7\% \cdot 342.72 = x$

$\quad 0.07 \cdot 342.72 = x$

$\quad\quad 23.9904 = x$

7% of \$342.72 is about \$23.99.

99. $32.16 = 8\% \cdot x$

$32.16 = 0.08 \cdot x$

$\dfrac{32.16}{0.08} = x$

$402 = x$

$32.16 is 8\% of $402.

101. $3\dfrac{1}{7} \div 2\dfrac{1}{2} = \dfrac{22}{7} \div \dfrac{5}{2}$

$= \dfrac{22}{7} \cdot \dfrac{2}{5}$

$= \dfrac{22 \cdot 2}{7 \cdot 5}$

$= \dfrac{44}{35}$

$= 1\dfrac{9}{35}$

103. $3x + 2 = 11$

$3x = 9$

$x = 3$

STUDY SET Section 8.5

VOCABULARY

1. In the American system, temperatures are measured in degrees <u>Fahrenheit</u>.

CONCEPTS

3. The formula used for changing degrees Celsius to degrees Fahrenheit is $F = \dfrac{9}{5}C + 32$.

NOTATION

5. 4,500 ft = 4,500(<u>0.3048</u> m)

$= \underline{1,371.6}$ m

$= 1.3716$ km

7. 8 L = 8(<u>0.264</u> gal)

$= 2.112$ gal

PRACTICE

9. 3 ft = 3(0.3048 m)

$= 0.9144$ m

≈ 91.4 cm

11. 3.75 m = 3.75(3.2808 ft)

$= 3.75(3.2808)(12$ in.$)$

≈ 147.6 in.

13. 12 km = 12(0.6214 mi)

$= 12(0.6214)(5,280$ ft$)$

$= 39,371.904$ ft

$\approx 39,372$ ft

15. 5,000 in.= 5,000(2.54 cm)

$= 12,700$ cm

$= 127$ m

17. 37 oz = 37(28.35 g)

$= 1,048.95$ g

$= 1.04895$ kg

≈ 1 kg

19. 25 lb = 25(0.454 kg)

$= 11.35$ kg

$= 11,350$ g

21. 0.5 kg = 0.5(2.2 lb)

$= 0.5(2.2)(16$ oz$)$

$= 17.6$ oz

23. 17 g = 17(0.035 oz)

$= 0.595$

≈ 0.6 oz

25. $3 \text{ fl oz} = 3(0.030 \text{ L})$
$\qquad = 0.090 \text{ L}$
$\qquad \approx 0.1 \text{ L}$

27. $7.2 \text{ L} = 7.2(33.8 \text{ fl oz})$
$\qquad = 243.36 \text{ fl oz}$
$\qquad \approx 243.4 \text{ fl oz}$

29. $0.75 \text{ qt} = 0.75(0.946 \text{ L})$
$\qquad = 0.7095 \text{ L}$
$\qquad = 709.5 \text{ mL}$
$\qquad \approx 710 \text{ mL}$

31. $500 \text{ mL} = 0.5 \text{ L}$
$\qquad = 0.5(1.06 \text{ qt})$
$\qquad = 0.53 \text{ qt}$
$\qquad \approx 0.5 \text{ qt}$

33. $C = \dfrac{5F - 160}{9}$

$\qquad = \dfrac{5(50) - 160}{9}$

$\qquad = \dfrac{250 - 160}{9}$

$\qquad = \dfrac{90}{9}$

$\qquad = 10° \text{ C}$

35. $F = \dfrac{9}{5}C + 32$

$\qquad = \dfrac{9}{5}(50) + 32$

$\qquad = 9 \cdot 10 + 32$

$\qquad = 90 + 32$

$\qquad = 122° \text{ F}$

37. $F = \dfrac{9}{5}C + 32$

$\qquad = \dfrac{9}{5}(-10) + 32$

$\qquad = 9(-2) + 32$

$\qquad = -18 + 32$

$\qquad = 14° \text{ F}$

39. $C = \dfrac{5F - 160}{9}$

$\qquad = \dfrac{5(-5) - 160}{9}$

$\qquad = \dfrac{-25 - 160}{9}$

$\qquad = \dfrac{-185}{9}$

$\qquad \approx -20.6° \text{ C}$

APPLICATIONS

41. Convert 8 kilometers to miles.
$\qquad 8 \text{ km} = 8(0.6214 \text{ mi})$
$\qquad \qquad = 4.9712 \text{ mi}$
$\qquad \qquad \approx 5 \text{ mi}$
The distance between the cities is about 5 miles.

43. Convert 90 miles to kilometers.
$\qquad 90 \text{ mi} = 90(1.6093 \text{ km})$
$\qquad \qquad = 144.837 \text{ km}$
$\qquad \qquad \approx 144.8 \text{ km}$
The distance between the cities is about 144.8 km.

45. Convert 112 kilometers to miles.
$\qquad 112 \text{ km} = 112(0.6214 \text{ mi})$
$\qquad \qquad = 69.5968 \text{ mi}$
$\qquad \qquad \approx 70 \text{ mi}$
A cheetah can run about 70 miles per hour.

47. Convert 7.264 kilograms to pounds.

7.264 kg = 7.264(2.2 lb)
$$= 15.9808 \text{ lb}$$
$$\approx 16 \text{ lb}$$

A shot-put weighs about 16 pounds.

49. Convert 13 kilograms to pounds.

13 kg = 13(2.2 lb)
$$= 28.6 \text{ lb}$$

The 30-lb box is heavier.

51. The unit cost of the 3-quart package is $\dfrac{\$4.50}{3 \text{ quarts}} = \$1.50$ per quart.

Convert 2 liters to quarts.

2 L = 2(1.06 qt)
$$= 2.12 \text{ qt}$$

The unit cost of the 2-liter package is

$$\frac{\$3.60}{2 \text{ liters}} \approx \frac{\$3.60}{2.12 \text{ quarts}}$$

$$\approx \$1.70 \text{ per quart.}$$

The 3-quart package is the better buy.

53. Convert the temperatures to degrees Fahrenheit. Convert 15°C to degrees Fahrenheit.

$$F = \frac{9}{5}C + 32$$

$$= \frac{9}{5}(15) + 32$$

$$= 9 \cdot 3 + 32$$

$$= 27 + 32$$

$$= 59° \text{ F}$$

Convert 28°C to degrees Fahrenheit.

$$F = \frac{9}{5}C + 32$$

$$= \frac{9}{5}(28) + 32$$

$$= \frac{9 \cdot 28}{5} + 32$$

$$= 50.4 + 32$$

$$= 82.4°\text{F}$$

Convert 50°C to degrees Fahrenheit.

$$F = \frac{9}{5}C + 32$$

$$= \frac{9}{5}(50) + 32$$

$$= 9 \cdot 10 + 32$$

$$= 90 + 32$$

$$= 122°\text{F}$$

For a shower, 59°F is somewhat cold, 122°F is too hot, and 82.4°F is comfortable. For a shower, 28°C is the best of the three temperatures given.

55. Convert the temperatures to degrees Fahrenheit. Convert $-5°C$ to degrees Fahrenheit.

$$F = \frac{9}{5}C + 32$$

$$= \frac{9}{5}(-5) + 32$$

$$= 9(-1) + 32$$

$$= -9 + 32$$

$$= 23°\,F$$

Convert $0°C$ to degrees Fahrenheit.

$$F = \frac{9}{5}C + 32$$

$$= \frac{9}{5}(0) + 32$$

$$= 32°\,F$$

Convert $10°C$ to degrees Fahrenheit.

$$F = \frac{9}{5}C + 32$$

$$= \frac{9}{5}(10) + 32$$

$$= 9 \bullet 2 + 32$$

$$= 18 + 32$$

$$= 50°\,F$$

It might snow if the temperature is 23°F or 32°F, but not 50°F. The temperatures at which it might snow are $-5°C$ and $0°C$.

REVIEW

59. $\begin{aligned} 6y + 7 - y - 3 &= (6y - y) + (7 - 3) \\ &= 5y + 4 \end{aligned}$

61. $\begin{aligned} -3(x - 4) - 2(2x + 6) &= -3(x) - 3(-4) - 2(2x) - 2(6) \\ &= -3x + 12 - 4x - 12 \\ &= (-3x - 4x) + (12 - 12) \\ &= -7x \end{aligned}$

63. $x \cdot x \cdot x = x^{1+1+1}$
$$= x^3$$

65. $3b(5b) = (3 \cdot 5)(b \cdot b)$
$$= 15b^2$$

CHAPTER 8 REVIEW

1. a. $\dfrac{4 \text{ inches}}{12 \text{ inches}} = \dfrac{1}{3}$

   b. $\dfrac{8 \text{ ounces}}{2 \text{ pounds}} = \dfrac{8 \text{ ounces}}{2(16 \text{ ounces})}$
   $$= \dfrac{8 \text{ ounces}}{32 \text{ ounces}}$$
   $$= \dfrac{1}{4}$$

3. $\dfrac{\$333.25}{43 \text{ hours}} = \$7.75 \text{ per hour}$

5. a. The fourth term is 75.

   b. The second term is 15.

7. a. $\dfrac{5}{9} \stackrel{?}{=} \dfrac{20}{36}$

   $5 \cdot 36 \stackrel{?}{=} 9 \cdot 20$

   $180 = 180$
   The numbers are proportional.

   b. $\dfrac{7}{13} \stackrel{?}{=} \dfrac{29}{54}$

   $7 \cdot 54 \stackrel{?}{=} 13 \cdot 29$

   $378 \neq 377$
   The numbers are not proportional.

9. 35 miles is to 2 gallons as m miles is to 11 gallons.
   $$\dfrac{35}{2} = \dfrac{m}{11}$$
   $$35 \cdot 11 = 2m$$
   $$385 = 2m$$
   $$\dfrac{385}{2} = m$$
   $192.5 = m$
   The truck can go 192.5 miles on 11 gallons of gas.

11. The mouse is about $1\dfrac{1}{2}$ in. long.

13. a. $32 \text{ oz} = 32 \text{ oz} \cdot \dfrac{1 \text{ lb}}{16 \text{ oz}}$
   $$= \dfrac{32}{16} \text{ lb}$$
   $$= 2 \text{ lb}$$

13. b. $17.2 \text{ lb} = 17.2 \text{ lb} \cdot \dfrac{16 \text{ oz}}{1 \text{ lb}}$

$= 17.2 \cdot 16 \text{ oz}$

$= 275.2 \text{ oz}$

c. $3 \text{ tons} = 3 \text{ tons} \cdot \dfrac{2,000 \text{ lb}}{1 \text{ ton}} \cdot \dfrac{16 \text{ oz}}{1 \text{ lb}}$

$= 3 \cdot 2,000 \cdot 16 \text{ oz}$

$= 96,000 \text{ oz}$

d. $4,500 \text{ lb} = 4,500 \text{ lb} \cdot \dfrac{1 \text{ ton}}{2,000 \text{ lb}}$

$= \dfrac{4,500}{2,000} \text{ tons}$

$= 2.25 \text{ tons}$

15 a. $20 \text{ min} = 20 \text{ min} \cdot \dfrac{60 \text{ sec}}{1 \text{ min}}$

$= 20 \cdot 60 \text{ sec}$

$= 1,200 \text{ sec}$

b. $900 \text{ sec} = 900 \text{ sec} \cdot \dfrac{1 \text{ min}}{60 \text{ sec}}$

$= \dfrac{900}{60} \text{ min}$

$= 15 \text{ min}$

c. $200 \text{ hr} = 200 \text{ hr} \cdot \dfrac{1 \text{ day}}{24 \text{ hr}}$

$= \dfrac{200}{24} \text{ days}$

$= \dfrac{25}{3} \text{ days}$

$= 8\dfrac{1}{3} \text{ days}$

15. d.
$$6 \text{ hr} = 6 \text{ hr} \cdot \frac{60 \text{ min}}{1 \text{ hr}}$$
$$= 6 \cdot 60 \text{ min}$$
$$= 360 \text{ min}$$

e.
$$4.5 \text{ days} = 4.5 \text{ days} \cdot \frac{24 \text{ hr}}{1 \text{ day}}$$
$$= 4.5 \cdot 24 \text{ hr}$$
$$= 108 \text{ hr}$$

f.
$$1 \text{ day} = 1 \text{ day} \cdot \frac{24 \text{ hr}}{1 \text{ day}} \cdot \frac{60 \text{ min}}{1 \text{ hr}} \cdot \frac{60 \text{ sec}}{1 \text{ hr}}$$
$$= 1 \cdot 24 \cdot 60 \cdot 60 \text{ sec}$$
$$= 86,400 \text{ sec}$$

17. Convert 50 gallons to magnums, using the unit conversion factor $\dfrac{1 \text{ magnum}}{2 \text{ quarts}} = 1.$

$$50 \text{ gal} = 50 \text{ gal} \cdot \frac{4 \text{ qt}}{1 \text{ gal}} \cdot \frac{1 \text{ magnum}}{2 \text{ qt}}$$
$$= \frac{50 \cdot 4}{2} \text{ magnums}$$
$$= 100 \text{ magnums}$$

19. a.
$$475 \text{ cm} = 475 \text{ cm} \cdot \frac{1 \text{ m}}{100 \text{ cm}}$$
$$= \frac{475}{100} \text{ m}$$
$$= 4.75 \text{ m}$$

b.
$$8 \text{ m} = 8 \text{ m} \cdot \frac{1,000 \text{ mm}}{1 \text{ m}}$$
$$= 8 \cdot 1,000 \text{ mm}$$
$$= 8,000 \text{ mm}$$

19. c. $3 \text{ dam} = 3 \text{ dam} \cdot \dfrac{10 \text{ m}}{1 \text{ dam}} \cdot \dfrac{1 \text{ km}}{1,000 \text{ m}}$

$\phantom{19. c.\ 3 \text{ dam}} = \dfrac{3 \cdot 10}{1,000} \text{ km}$

$\phantom{19. c.\ 3 \text{ dam}} = 0.03 \text{ km}$

d. $2 \text{ hm} = 2 \text{ hm} \cdot \dfrac{100 \text{ m}}{1 \text{ hm}} \cdot \dfrac{10 \text{ dm}}{1 \text{ m}}$

$\phantom{\text{d.}\ 2 \text{ hm}} = 2 \cdot 100 \cdot 10 \text{ dm}$

$\phantom{\text{d.}\ 2 \text{ hm}} = 2,000 \text{ dm}$

e. $5 \text{ km} = 5 \text{ km} \cdot \dfrac{1,000 \text{ m}}{1 \text{ km}} \cdot \dfrac{1 \text{ hm}}{100 \text{ m}}$

$\phantom{\text{e.}\ 5 \text{ km}} = \dfrac{5 \cdot 1,000}{100} \text{ hm}$

$\phantom{\text{e.}\ 5 \text{ km}} = 50 \text{ hm}$

f. $2,500 \text{ m} = 2,500 \text{ m} \cdot \dfrac{1 \text{ hm}}{100 \text{ m}}$

$\phantom{\text{f.}\ 2,500 \text{ m}} = \dfrac{2,500}{100} \text{ hm}$

$\phantom{\text{f.}\ 2,500 \text{ m}} = 25 \text{ hm}$

21. The 100 caplets contain
$100 \cdot 500$ milligrams = 50,000 milligrams of Tylenol.
Convert 50,000 milligrams to grams.

$50,000 \text{ mg} = 50,000 \text{ mg} \cdot \dfrac{1 \text{ g}}{1,000 \text{ mg}}$

$\phantom{50,000 \text{ mg}} = \dfrac{50,000}{1,000} \text{ g}$

$\phantom{50,000 \text{ mg}} = 50 \text{ g}$

The bottle contains 50 grams of Tylenol.

23. Convert 2 liters to centiliters.

$$2\text{ L} = 2\text{ L} \cdot \frac{100\text{ cL}}{1\text{ L}}$$

$$= 2 \cdot 100\text{ cL}$$

$$= 200\text{ cL}$$

25. Convert 419 meters to feet.

419 m = 419(3.2808 ft)

= 1,374.6552 ft

So 419 meters is more than 1,250 feet and the World Trade Center is taller.

27. Convert 6 feet, 6 inches to centimeters.

6 ft 6 in. = 6 ft + 6 in.

= 6(12 in.) + 6 in.

= 72 in. + 6 in.

= 78 in.

78 in. = 78(2.54 cm)

= 198.12 cm

Michael Jordan is 198.12 cm tall.

29. Convert 84 kilograms to pounds.

84 kg = 84(2.2 lb)

= 184.8 lb

The man weighs 184.8 pounds.

31. The unit cost for the 1-gallon bottle is $\dfrac{\$1.39}{1\text{ gallon}} = \$1.39$ per gallon.

Convert 5 liters to gallons.

5 L = 5(0.264 gal)

= 1.32 gal

The unit cost for the 5-liter bottle is

$$\frac{\$1.80}{5\text{ liters}} \approx \frac{\$1.80}{1.32\text{ gallons}}$$

$$\approx \$1.36 \text{ per gallon}$$

The 5-liter bottle is the better buy.

33.

$$C = \frac{5F - 160}{9}$$

$$= \frac{5(77) - 160}{9}$$

$$= \frac{385 - 160}{9}$$

$$= \frac{225}{9}$$

$$= 25°\text{C}$$

35. Convert the temperatures to degrees Fahrenheit. Convert 10°C to degrees Fahrenheit.

$$F = \frac{9}{5}C + 32$$

$$= \frac{9}{5}(10) + 32$$

$$= 9 \cdot 2 + 32$$

$$= 18 + 32$$

$$= 50°\,F$$

Convert 30°C to degrees Fahrenheit.

$$F = \frac{9}{5}C + 32$$

$$= \frac{9}{5}(30) + 32$$

$$= 9 \cdot 6 + 32$$

$$= 54 + 32$$

$$= 86°\,F$$

Convert 50°C to degrees Fahrenheit.

$$F = \frac{9}{5}C + 32$$

$$= \frac{9}{5}(50) + 32$$

$$= 9 \cdot 10 + 32$$

$$= 90 + 32$$

$$= 122°\,F$$

Water that is 122°F is too hot for swimming, so hotter water would also be too hot for swimming. Swimming would be comfortable in water that was 86°F while 50°F is too cold for swimming. Of the temperatures given, 30°C would be comfortable for swimming.

KEY CONCEPT Proportions

1. ***Step 1:***
Let x = the number of <u>teacher's aides needed to supervise 75 children</u>. 15 children are to <u>2</u> aides as <u>75</u> children are to <u>x</u> aides. Expressing this as a proportion, we have

$$\frac{15}{2} = \frac{75}{x}.$$

Step 2:
Solve for x: $\dfrac{15}{2} = \dfrac{75}{x}$

$$\underline{15} \cdot x = 2 \cdot \underline{75}$$

$$15x = 150$$

$$\frac{15x}{\underline{15}} = \frac{150}{\underline{15}}$$

$$x = \underline{10}$$

Ten teacher's aides are needed to supervise 75 children.

3. 2 seconds are to 3 feet of film as 120 minutes are to x feet of film. First convert 120 minutes into seconds.

$$120 \text{ min} = 120 \text{ min} \cdot \frac{60 \text{ sec}}{1 \text{ min}}$$

$$= 120 \cdot 60 \text{ sec}$$

$$= 7,200 \text{ sec}$$

$$\frac{2 \text{ sec}}{3 \text{ ft}} = \frac{120 \text{ min}}{x \text{ ft}}$$

$$\frac{2 \text{ sec}}{3 \text{ ft}} = \frac{7,200 \text{ sec}}{x \text{ ft}}$$

$$\frac{2}{3} = \frac{7,200}{x}$$

$$2 \cdot x = 3 \cdot 7,200$$

$$2x = 21,600$$

$$\frac{2x}{2} = \frac{21,600}{2}$$

$$x = 10,800$$

There are 10,800 feet of film in a 120-minute movie.

CHAPTER 8 TEST

1. $\dfrac{6 \text{ ft}}{8 \text{ ft}} = \dfrac{3 \cdot 2}{4 \cdot 2}$

$$= \frac{3}{4}$$

3. The unit cost for the two-pound can is

$$\frac{\$3.38}{2 \text{ pounds}} = \$1.69 \text{ per pound.}$$

The unit cost for the five-pound can is

$$\frac{\$8.50}{5 \text{ pounds}} = \$1.70 \text{ per pound.}$$

The two-pound can is the better buy.

5. The product of the extremes is $25 \cdot 460 = 11,500$. The product of the means is $33 \cdot 350 = 11,550$. Since the products are not equal, the statement is not a proportion.

7. The third term of the proportion is c.

9. $\dfrac{x}{3} = \dfrac{35}{7}$

$$x \cdot 7 = 3 \cdot 35$$

$$7x = 105$$

$$x = 15$$

11. $\dfrac{2x+3}{5} = \dfrac{5}{1}$

$$(2x+3)1 = 5 \cdot 5$$

$$2x + 3 = 25$$

$$2x = 22$$

$$x = 11$$

13. 13 ounces is to $\$2.79$ as 16 ounces is to $\$x$.

$$\frac{13}{2.79} = \frac{16}{x}$$

$$13 \cdot x = 2.79 \cdot 16$$

$$13x = 44.64$$

$$x = \frac{44.64}{13}$$

$$x \approx 3.43$$

You would expect to pay $\$3.43$ for 16 ounces.

15. $180 \text{ in.} = 180 \text{ in.} \cdot \dfrac{1 \text{ ft}}{12 \text{ in.}}$

$$= \frac{180}{12} \text{ ft}$$

$$= 15 \text{ ft}$$

17. $10 \text{ lb} = 10 \text{ lb} \cdot \dfrac{16 \text{ oz}}{1 \text{ lb}}$

$\qquad = 10 \cdot 16 \text{ oz}$

$\qquad = 160 \text{ oz}$

19. Convert 1 gallon to fluid ounces.

$1 \text{ gal} = 1 \text{ gal} \cdot \dfrac{4 \text{ qt}}{1 \text{ gal}} \cdot \dfrac{2 \text{ pt}}{1 \text{ qt}} \cdot \dfrac{2 \text{ c}}{1 \text{ pt}} \cdot \dfrac{8 \text{ fl oz}}{1 \text{ c}}$

$\qquad = 1 \cdot 4 \cdot 2 \cdot 2 \cdot 8 \text{ fl oz}$

$\qquad = 128 \text{ fl oz}$

There are 128 fluid ounces of milk in a 1-gallon carton.

21. Convert 5 meters to centimeters.

$5 \text{ m} = 5 \text{ m} \cdot \dfrac{100 \text{ cm}}{1 \text{ m}}$

$\qquad = 5 \cdot 100 \text{ cm}$

$\qquad = 500 \text{ cm}$

There are 500 centimeters in 5 meters.

23. Convert 10 kilometers to meters.

$10 \text{ km} = 10 \text{ km} \cdot \dfrac{1{,}000 \text{ m}}{1 \text{ km}}$

$\qquad = 10 \cdot 1{,}000 \text{ m}$

$\qquad = 10{,}000 \text{ m}$

There are 10,000 meters in 10 kilometers.

25. $70 \text{ L} = 70 \text{ L} \cdot \dfrac{1 \text{ hL}}{100 \text{ L}}$

$\qquad = \dfrac{70}{100} \text{ hL}$

$\qquad = 0.7 \text{ hL}$

27. Convert 100 yards to meters.
$100 \text{ yd} = 100(0.9144 \text{ m})$

$\qquad = 91.44 \text{ m}$

The 100-yard race is longer than the 80-meter race.

29. The unit cost of the 1-liter bottle is $\dfrac{\$0.89}{1 \text{ liter}} = \$0.89$ per liter.

Convert 2 quarts to liters.

$2 \text{ qt} = 2(0.946 \text{ L})$

$\qquad = 1.892 \text{ L}$

The unit cost of the 2-quart bottle is

$$\frac{\$1.73}{2 \text{ quarts}} \approx \frac{\$1.73}{1.892 \text{ liters}}$$

$$\approx \$0.91 \text{ per liter}$$

The 1-liter bottle is the better buy.

STUDY SET Section 9.1

VOCABULARY

1. A line <u>segment</u> has two endpoints.

3. A <u>midpoint</u> divides a line segment into two parts of equal length.

5. A <u>protractor</u> is used to measure angles.

7. A <u>right</u> angle measures 90°.

9. The measure of a straight angle is <u>180°</u>.

11. The sum of two <u>supplementary</u> angles is 180°.

CONCEPTS

13. True

15. False; lines do not have endpoints.

17. True; both angles have measure 30°. (They are also vertical angles.)

19. True, these are vertical angles.

21. $m(\angle AGC) = 30° < 90°$, so $\angle AGC$ is an acute angle.

23. $m(\angle FGD) = 90° + 30° = 120° > 90°$, so $\angle FGD$ is an obtuse angle.

25. $m(\angle BGE) = 30° + 60° = 90°$, so $\angle BGE$ is a right angle.

27. $m(\angle DGC) = 60° + 90° + 30° = 180°$, so $\angle DGC$ is a straight angle.

29. True

31. False; using a protractor, $m(\angle AGB) = 102°$, while $m(\angle BGC) = 40°$.

33. Yes, $\angle 1$ and $\angle 2$ are vertical angles so they are congruent.

35. Yes, $\angle AGB$ and $\angle DGE$ are vertical angles so they are congruent.

37. No, using a protractor, $m(\angle AGF) = 60°$ while $m(\angle FGE) = 30°$.

39. True

41. True

43. True; $m(\angle AGE) = 90°$.

45. True, since points E, G, and B all lie on one line.

NOTATION

47. The symbol $\angle$ means <u>angle</u>.

49. The symbol $\overrightarrow{AB}$ is read as "<u>ray</u> AB."

PRACTICE

51. $m\left(\overline{AC}\right) = 5 - 2$
$= 3$

53. $m\left(\overline{CE}\right) = 8 - 5$
$= 3$

55. $m\left(\overline{CD}\right) = 6 - 5$
$= 1$

57. $m(\overline{AD}) = 6 - 2$
$= 4$
$m(\overline{AB}) = 4 - 2$
$= 2$
$m(\overline{BD}) = 6 - 4$
$= 2$
B is the midpoint of $\overline{AD}$.

59. $40°$

61. $135°$

63. $x + 45 = 55$
$x = 10$

65. $x + 22.5 = 50$
$x = 27.5$

67. $2x = x + 30$
$x = 30$

69. $4x + 15 = 7x - 60$
$15 = 3x - 60$
$75 = 3x$
$25 = x$

71.

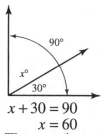

$x + 30 = 90$
$x = 60$
The complement of a $30°$ angle is a $60°$ angle.

73.

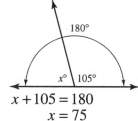

$x + 105 = 180$
$x = 75$
The supplement of a $105°$ angle is a $75°$ angle.

75. $\angle 1$ and $\angle 4$ are supplementary angles so
$m(\angle 1) + m(\angle 4) = 180°.$
$50° + m(\angle 4) = 180°$
$m(\angle 4) = 130°$

77. $\angle 1$ and $\angle 3$ are vertical angles, as are $\angle 2$ and $\angle 4$, so $m(\angle 1) = m(\angle 3)$ and $m(\angle 2) = m(\angle 4)$.
$m(\angle 1) = m(\angle 3)$
$50° = m(\angle 3)$
From Exercise 75, $m(\angle 4) = 130°$.
$m(\angle 2) = m(\angle 4)$
$m(\angle 2) = 130°$
$m(\angle 1) + m(\angle 2) + m(\angle 3) = 50° + 130° + 50°$
$= 180° + 50°$
$= 230°$

79. ∠1 and the 100° angle are vertical angles, so m(∠1) = 100°.

81. ∠1 and the 100° angle are vertical angles, so m(∠1) = 100°.
∠3 ≅ ∠4, so m(∠3) = m(∠4).

$$m(\angle 1) + m(\angle 3) + m(\angle 4) = 180°$$
$$100° + m(\angle 3) + m(\angle 3) = 180°$$
$$2 \cdot m(\angle 3) = 80°$$
$$m(\angle 3) = 40°$$

APPLICATIONS

83. Vertical or horizontal blinds on windows are lines. Answers will vary.

85. The angle that the open hood of a car makes with the car is an acute angle. Answers will vary.

87. a. 80°

b. 30°

c. 65°

REVIEW

91.
$$2^4 = 2 \cdot 2 \cdot 2 \cdot 2$$
$$= 4 \cdot 2 \cdot 2$$
$$= 8 \cdot 2$$
$$= 16$$

93.
$$\frac{3}{4} - \frac{1}{8} - \frac{1}{3} = \frac{3 \cdot 6}{4 \cdot 6} - \frac{1 \cdot 3}{8 \cdot 3} - \frac{1 \cdot 8}{3 \cdot 8}$$
$$= \frac{18}{24} - \frac{3}{24} - \frac{8}{24}$$
$$= \frac{18 - 3 - 8}{24}$$
$$= \frac{15 - 8}{24}$$
$$= \frac{7}{24}$$

95.
$$\frac{12}{17} \div \frac{4}{34} = \frac{12}{17} \cdot \frac{34}{4}$$
$$= \frac{3 \cdot 4 \cdot 2 \cdot 17}{17 \cdot 4}$$
$$= 3 \cdot 2$$
$$= 6$$

97.
$$5 \cdot 3 + 4 \cdot 2 = 15 + 8$$
$$= 23$$

STUDY SET Section 9.2
VOCABULARY

1. Two lines in the same plane are coplanar.

3. If two lines intersect and form right angles, they are perpendicular.

5. In Illustration 1, ∠4 an ∠6 are alternate interior angles.

CONCEPTS

7. ∠4 and ∠6 are alternate interior angles as are ∠3 and ∠5.

9. ∠3, ∠4, ∠5, and ∠6 are interior angles.

11. l_1 and l_3 are both parallel to l_2, so they are parallel to each other.

NOTATION

13. The symbol ⌐ indicates <u>a right angle</u>.

15. The symbol ⊥ is read as "<u>is perpendicular to</u>."

PRACTICE

17. $m(\angle 1) = 130°$ since $\angle 1$ and $\angle 4$ are vertical angles.

$m(\angle 2) = 50°$ since $\angle 2$ and $\angle 4$ are supplementary angles.

$m(\angle 3) = 50°$ since $\angle 3$ and $\angle 2$ are vertical angles.

$m(\angle 5) = 130°$ since $\angle 5$ and $\angle 4$ are alternate interior angles and $l_1 \parallel l_2$.

$m(\angle 6) = 50°$ since $\angle 6$ and $\angle 5$ are supplementary angles.

$m(\angle 7) = 50°$ since $\angle 7$ and $\angle 6$ are vertical angles.

$m(\angle 8) = 130°$ since $\angle 8$ and $\angle 5$ are vertical angles.
(Other reasons can be used to find the angles.)

19. $m(\angle A) = 50°$ since $\angle A$ and the $50°$ angle are alternate interior angles and $l_1 \parallel \overline{AB}$.

$m(\angle 3) = 135°$ since $\angle 3$ and the $45°$ angle are supplementary angles.

$m(\angle 2) = 45°$ since $\angle 2$ and the $45°$ angle are alternate interior angles and $l_1 \parallel \overline{AB}$.

$\angle 1$, $\angle 2$, and the $50°$ angle combine to form a straight angle.
$$m(\angle 1) + m(\angle 2) + 50° = 180°$$
$$m(\angle 1) + 45° + 50° = 180°$$
$$m(\angle 1) + 95° = 180°$$
$$m(\angle 1) = 85°$$

21. Since $l_1 \parallel l_2$, the $(5x)°$ angle and the $(6x - 10)°$ angle are congruent alternate interior angles.
$$5x = 6x - 10$$
$$0 = x - 10$$
$$10 = x$$

23. The $(2x + 10)°$ angle and the $(4x - 10)°$ angle are interior angles on the same side of the transversal, so they are supplementary.
$$(2x + 10) + (4x - 10) = 180$$
$$(2x + 4x) + (10 - 10) = 180$$
$$6x = 180$$
$$x = 30$$

25. The $x°$ angle and the $(3x + 20)°$ angle are interior angles on the same side of the transversal $\overline{BC}$, they are supplementary.
$$x + (3x + 20) = 180$$
$$(x + 3x) + 20 = 180$$
$$4x + 20 = 180$$
$$4x = 160$$
$$x = 40$$

27. The $(6x - 2)°$ angle and the $(9x - 38)°$ angle are congruent alternate interior angles created by the transversal $\overline{AE}$.
$$6x - 2 = 9x - 38$$
$$-2 = 3x - 38$$
$$36 = 3x$$
$$12 = x$$

APPLICATIONS

29. The rafters supporting one section of a roof are parallel. Answers will vary.

31. The beams for the walls will be perpendicular to the beams for the floors. Answers will vary.

33. a. A pair of scissors shows intersecting lines and vertical angles.

b. A rake shows parallel and perpendicular lines.

REVIEW

39. 60% of 120 is what number?
$$60\% \bullet 120 = x$$
$$0.60 \bullet 120 = x$$
$$72 = x$$
60% of 120 is 72.

41. $x \bullet 500 = 225$
$$x = \frac{225}{500}$$
$$x = 0.45$$
$$x = 45\%$$
225 is 45% of 500.

43. Yes, every whole number is an integer.

STUDY SET Section 9.3

VOCABULARY

1. A polygon with four sides is called a quadrilateral.

3. A hexagon is a polygon with six sides.

5. An eight-sided polygon is an octagon.

7. A triangle with three sides of equal length is called an equilateral triangle.

9. The longest side of a right triangle is the hypotenuse.

11. A <u>parallelogram</u> with a right <u>angle</u> is a rectangle.

13. A <u>rhombus</u> is a parallelogram with four sides of <u>equal</u> length.

15. The legs of an <u>isosceles</u> trapezoid have the same <u>length</u>.

CONCEPTS

17. The polygon has 4 sides so it is a quadrilateral. It has 4 vertices.

19. The polygon has 3 sides so it is a triangle. It has 3 vertices.

21. The polygon has 5 sides so it is a pentagon. It has 5 vertices.

23. The polygon has 6 sides so it is a hexagon.
It has 6 vertices.

25. No sides have equal length, so the triangle is a scalene triangle.

27. One angle is a right angle so the triangle is a right triangle. It is also a scalene triangle because no sides have equal length.

29. All three angles have equal measure so the triangle is an equilateral triangle.

31. Two sides have equal length so the triangle is an isosceles triangle.

33. All sides have equal length and all angles are right angles so the quadrilateral is a square.

35. This is a parallelogram with sides of equal length so the quadrilateral is a rhombus.

37. This is a parallelogram with four right angles so the quadrilateral is a rectangle.

39. Exactly two sides are parallel so the quadrilateral is a trapezoid.

NOTATION

41. The symbol $\triangle$ means <u>triangle</u>.

43.
$$m(\angle A) + m(\angle B) + m(\angle C) = 180°$$
$$30° + 60° + m(\angle C) = 180°$$
$$90° + m(\angle C) = 180°$$
$$m(\angle C) = 90°$$

45.
$$m(\angle A) + m(\angle B) + m(\angle C) = 180°$$
$$35° + 100° + m(\angle C) = 180°$$
$$135° + m(\angle C) = 180°$$
$$m(\angle C) = 45°$$

47.
$$m(\angle A) + m(\angle B) + m(\angle C) = 180°$$
$$25.5° + 63.8° + m(\angle C) = 180°$$
$$89.3° + m(\angle C) = 180°$$
$$m(\angle C) = 90.7°$$

49. $m(\angle 1) = 30°$ since $\angle 1$ and the $60°$ angle are complementary.

51. $m(\angle 2) = 60°$ since $\angle 2$ and the $60°$ angle are alternate interior angles created by the transversal $\overline{AC}$.

53.
$$S = (n-2)180°$$
$$= (6-2)180°$$
$$= (4)180°$$
$$= 720°$$

55.
$$S = (n-2)180°$$
$$= (10-2)180°$$
$$= (8)180°$$
$$= 1,440$$

57.
$$S = (n-2)180°$$
$$900° = (n-2)180°$$
$$\frac{900°}{180°} = n-2$$
$$5 = n-2$$
$$7 = n$$
A 7-sided polygon

59.
$$S = (n-2)180°$$
$$2{,}160° = (n-2)180°$$
$$\frac{2{,}160°}{180°} = n-2$$
$$12 = n-2$$
$$14 = n$$
A 14-sided polygon

APPLICATIONS

61. When you cut across a parking lot rather than going around a corner on the sidewalk, your path and the sidewalks form a triangle. Answers will vary.

63. Chess and checker boards are square and the red and black areas on them are squares.

REVIEW

67. What number is 20% of 110?
$$x = 20\% \cdot 110$$
$$x = 0.20 \cdot 110$$
$$x = 22$$
22 is 20% of 110.

69. What number is 20% of $\frac{6}{11}$?
$$x = 20\% \cdot \frac{6}{11}$$
$$x = \frac{20}{100} \cdot \frac{6}{11}$$
$$x = \frac{1}{5} \cdot \frac{6}{11}$$
$$x = \frac{6}{5 \cdot 11}$$
$$x = \frac{6}{55}$$
$\frac{6}{55}$ is 20% of $\frac{6}{11}$.

71.
$$x \cdot 200 = 80$$
$$x = \frac{80}{200}$$
$$x = 0.4$$
$$x = 40\%$$
80 is 40% of 200.

73.
$$20\% \cdot x = 500$$
$$0.20 \cdot x = 500$$
$$x = \frac{500}{0.20}$$
$$x = 2{,}500$$
20% of 2,500 is 500.

75.
$$0.85 \div 2(0.25) = 0.85 \div 2 \cdot 0.25$$
$$= \frac{85}{100} \div \frac{2}{1} \cdot 0.25$$
$$= \frac{85}{100} \cdot \frac{1}{2} \cdot 0.25$$
$$= \frac{85}{100} \cdot \frac{1}{2} \cdot \frac{25}{100}$$
$$= \frac{85 \cdot 25}{100 \cdot 2 \cdot 100}$$
$$= 0.10625$$

STUDY SET Section 9.4

VOCABULARY

1. <u>Congruent</u> triangles are the same size and the same shape.

3. If two triangles are <u>similar</u>, they have the <u>same</u> shape.

CONCEPTS

5. True; this is the SSS property.

7. False; if the congruent angle is between the pairs of sides on one triangle, but not on the other triangle, the triangles would not necessarily be congruent.

9. Yes; both triangles have sides of length 4 cm and 8 cm and the angle between these sides is a right angle in both triangles.

11. In a proportion, the <u>product</u> of the means is equal to the <u>product</u> of the <u>extremes</u>.

13. Yes, the triangles are similar since the angles at C are vertical angles, hence congruent. Thus two angles in the top triangle are congruent to two angles in the bottom triangle.

15. True

NOTATION

17. The symbol $\cong$ is read as "<u>is congruent to</u>."

PRACTICE

19. $\overline{AC}$ corresponds to $\overline{DF}$ since both sides have one slash.

$\overline{DE}$ corresponds to $\overline{AB}$ since both sides have two slashes.

$\overline{BC}$ corresponds to $\overline{EF}$ since both sides have three slashes.

$\angle A$ corresponds to $\angle D$ since these angles are between pairs of corresponding sides.

$\angle E$ corresponds to $\angle B$ since these angles are between pairs of corresponding sides.

$\angle F$ corresponds to $\angle C$ since these angles are between pairs of corresponding sides.

21. The triangles are congruent by the SSS property since three sides of the top triangle are congruent to three sides of the bottom triangle.

23. The triangles are not necessarily congruent since only one pair of sides and one pair of angles are congruent. (Note that we cannot assume the top and bottome segments are parallel.)

25. The triangles are congruent by the SSS property since the unmarked side is shared by both triangles.

27. The triangles are congruent by the SAS property since the right angles are between the sides of length 4 cm and the side shared by both triangles.

29. The triangles are congruent by the ASA property, so $x = 6$ cm.

31. The triangles are congruent by the SSS property, so $x = 50°$.

33. Two angles of the smaller triangle are congruent to two angles of the larger triangle so the triangles are similar.

35. $c^2 = a^2 + b^2$
$c^2 = 3^2 + 4^2$
$c^2 = 9 + 16$
$c^2 = 25$
$\sqrt{c^2} = \sqrt{25}$
$c = 5$

37. $a^2 + b^2 = c^2$
$15^2 + b^2 = 17^2$
$225 + b^2 = 289$
$b^2 = 64$
$\sqrt{b^2} = \sqrt{64}$
$b = 8$

39. $a^2 + b^2 = c^2$
$5^2 + b^2 = 9^2$
$25 + b^2 = 81$
$b^2 = 56$
$\sqrt{b^2} = \sqrt{56}$
$b = \sqrt{56}$

41. $17^2 \overset{?}{=} 8^2 + 15^2$
$289 \overset{?}{=} 64 + 225$
$289 = 289$

Since $17^2 = 8^2 + 15^2$, the triangle is a right triangle.

43. $26^2 \overset{?}{=} 7^2 + 24^2$
$676 \overset{?}{=} 49 + 576$
$676 \neq 625$

Since $26^2 \neq 7^2 + 24^2$, the triangle is not a right triangle.

APPLICATIONS

45. The triangles are similar so lengths of corresponding sides are in proportion.
$$\frac{h \text{ ft}}{6 \text{ ft}} = \frac{24 \text{ ft}}{4 \text{ ft}}$$
$h \cdot 4 = 6 \cdot 24$
$4h = 144$
$h = 36$
The tree is 36 feet tall.

47. The triangles are similar so lengths of corresponding sides are in proportion.
$$\frac{w \text{ ft}}{20 \text{ ft}} = \frac{74 \text{ ft}}{25 \text{ ft}}$$
$w \cdot 25 = 20 \cdot 74$
$25w = 1,480$
$w = 59.2$
The river is 59.2 feet wide.

49. As in Illustration 11, the triangles will be similar so corresponding sides will be in proportion.

$$\frac{x \text{ ft}}{1,200 \text{ ft}} = \frac{5 \text{ mi}}{1.5 \text{ mi}}$$

$$x \cdot 1.5 = 1,200 \cdot 5$$

$$1.5x = 6,000$$

$$x = 4,000$$

The plane will lose 4,000 ft in altitude.

51. The ladder, wall, and the ground will form a right triangle with the ladder as the hypotenuse. Use the Pythagorean theorem with $a = 16$, $c = 20$, and b the unknown quantity.

$$a^2 + b^2 = c^2$$

$$16^2 + b^2 = 20^2$$

$$256 + b^2 = 400$$

$$b^2 = 144$$

$$\sqrt{b^2} = \sqrt{144}$$

$$b = 12$$

53. The distance from home plate to second base is the length of the hypotenuse of a right triangle whose legs have length 90 feet. Use the Pythagorean theorem with $a = 90$, $b = 90$, and c the unknown quantity.

$$c^2 = a^2 + b^2$$

$$c^2 = 90^2 + 90^2$$

$$c^2 = 8,100 + 8,100$$

$$c^2 = 16,200$$

$$\sqrt{c^2} = \sqrt{16,200}$$

$$c \approx 127.3$$

The distance from home plate to second base is approximately 127.3 feet.

REVIEW

57. 0.95 is close to 1, 3.89 is close to 4, and 2.997 is close to 3. An estimation for $\dfrac{0.95 \cdot 3.89}{2.997}$ is

$$\frac{1 \cdot 4}{3} = \frac{4}{3}$$

$$= 1\frac{1}{3}$$

59. 32% is close to $33\frac{1}{3}$% which is $\frac{1}{3}$. An estimation for 32% of 60 is

$$\frac{1}{3} \cdot 60 = 20.$$

61.
$$2 + 4 \cdot 3^2 = 2 + 4 \cdot 9$$
$$= 2 + 36$$
$$= 38$$

STUDY SET Section 9.5

VOCABULARY

1. The distance around a polygon is called the <u>perimeter</u>.

3. The measure of the surface enclosed by a polygon is called its <u>area</u>.

5. The area of a polygon is measured in <u>square</u> units.

CONCEPTS

7. Answers may vary.

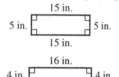

9.

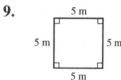

11. Answers may vary.

13. Answers may vary.

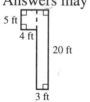

NOTATION

15. The formula for the perimeter of a square is <u>$P = 4s$</u>.

17. The symbol 1 in.^{2} means <u>one square inch</u>.

19. The formula for the area of a square is <u>$A = s^2$</u>.

21. The formula $A = \dfrac{1}{2}bh$ gives the area of a <u>triangle</u>.

PRACTICE

23. Since the figure is a square, the perimeter is given by $P = 4s$.
$$P = 4s$$
$$P = 4(8)$$
$$= 32$$
The perimeter of the square is 32 inches.

25. The width of the figure is 10 meters. The figure can be thought of as 3 stacked rectangles. The top and bottom rectangles both have width 4 meters, so the middle rectangle must have width 2 meters. The unlabeled side has length 2 meters.
$$P = 10 + 6 + 4 + 2 + 2 + 2 + 4 + 6$$
$$= 36$$
The perimeter of the figure is 36 meters.

27. $P = 6 + 6 + 8 + 10 + 7$
$$= 37$$
The perimeter of the figure is 37 centimeters.

29. $P = 21 + 32 + 32$
$$= 85$$
The perimeter of the triangle is 85 centimeters.

31. The sides of an equilateral triangle all have the same length.
Let s = the length of the sides of the triangle.

$$P = s + s + s$$
$$85 = s + s + s$$
$$85 = 3s$$
$$\frac{85}{3} = s$$
$$28\frac{1}{3} = s$$

Each side of the triangle has length $28\frac{1}{3}$ feet.

33. The figure is a square so the area is given by $A = s^2$.

$$A = s^2$$
$$A = 4^2$$
$$= 16$$

The area is 16 cm^2.

35. The figure is a parallelogram so the area is given by $A = bh$.

$$A = bh$$
$$A = (15)(4)$$
$$= 60$$

The area is 60 cm^2.

37. The figure is a triangle so the area is given by $A = \frac{1}{2}bh$.

$$A = \frac{1}{2}bh$$
$$A = \frac{1}{2}(10)(5)$$
$$= 25$$

The area is 25 in.^{2}.

39. The figure is a trapezoid, so the area is given by $A = \frac{1}{2}h(b_1 + b_2)$.

$$A = \frac{1}{2}h(b_1 + b_2)$$
$$A = \frac{1}{2}(13)(17 + 9)$$
$$= \frac{1}{2}(13)(26)$$
$$= 169$$

The area is 169 mm^2.

41. The figure is a combination of a square and a triangle. The square has sides of length 8 m.

$$A_{\text{square}} = s^2$$
$$= 8^2$$
$$= 64$$

The triangle has base of length 8 m and height of 4 m.

$$A_{\text{triangle}} = \frac{1}{2}bh$$
$$= \frac{1}{2}(8)(4)$$
$$= 16$$

The area of the figure is

$$A = A_{\text{square}} + A_{\text{triangle}}$$
$$= 64 + 16$$
$$= 80$$

The area is 80 m^2.

43. The shaded area is the area of a square with a triangle taken out. The square has sides of length 10 yd.

$$A_{square} = s^2$$
$$= 10^2$$
$$= 100$$

The triangle has base of length 10 yd and height of 5 yd.

$$A_{triangle} = \frac{1}{2}bh$$
$$= \frac{1}{2}(10)(5)$$
$$= 25$$

The area of the shaded figure is

$$A = A_{square} - A_{triangle}$$
$$= 100 - 25$$
$$= 75$$

The area is 75 yd^2.

45. The shaded area is the area of a rectangle with a square taken out. The rectangle has length 14 m and width 6 m.

$$A_{rectangle} = lw$$
$$= (14)(6)$$
$$= 84$$

The square has sides of length 3 ft.

$$A_{square} = s^2$$
$$= 3^2$$
$$= 9$$

The area of the shaded figure is

$$A = A_{rectangle} - A_{square}$$
$$= 84 - 9$$
$$= 75$$

The area is 75 m^2.

47. $1 \text{ ft}^2 = (1 \text{ ft})(1 \text{ ft})$
$$= (12 \text{ in.})(12 \text{ in.})$$
$$= 144 \text{ in.}^2$$

APPLICATIONS

49. The fencing will go around the perimeter of the yard.

$$P = 2l + 2w$$
$$P = 2(110) + 2(85)$$
$$= 220 + 170$$
$$= 390$$

The perimeter is 390 feet. The cost of 390 feet of fencing is $2.44(390) = \$951.60$.

51. Starting by the house, the woman plants one tree. 3 feet along the side of the yard, she plants another tree. Thus, there are 2 trees in the first 3 feet, 3 trees in the first 6 feet, and so on. The number of trees is one more than the number of feet divided by 3. The three sides of the yard have length 70 ft + 100 ft + 70 ft = 240 ft. The number of trees is

$$\frac{1}{3}(240) + 1 = 80 + 1$$
$$= 81$$

53. The unit cost of the ceramic tile is

$$\frac{\$3.75}{1 \text{ ft}^2} = \$3.75 \text{ per ft}^2.$$

$$1 \text{ yd}^2 = (1 \text{ yd})(1 \text{ yd})$$
$$= (3 \text{ ft})(3 \text{ ft})$$
$$= 9 \text{ ft}^2$$

The unit cost of the linoleum is

$$\frac{\$34.95}{1 \text{ yd}^2} = \frac{\$34.95}{9 \text{ ft}^2}$$
$$\approx \$3.88 \text{ per ft}^2$$

The linoleum is more expensive.

55. Convert 24 feet and 15 feet to yards.

$$24 \text{ ft} = 24 \text{ ft} \cdot \frac{1 \text{ yd}}{3 \text{ ft}}$$

$$= \frac{24}{3} \text{ yd}$$

$$= 8 \text{ yd}$$

$$15 \text{ ft} = 15 \text{ ft} \cdot \frac{1 \text{ yd}}{3 \text{ ft}}$$

$$= \frac{15}{3} \text{ yd}$$

$$= 5 \text{ yd}$$

The area of carpeting needed is the area of the rectangular room.

$A = lw$

$A = (24 \text{ ft})(15 \text{ ft})$

$ = (8 \text{ yd})(5 \text{ yd})$

$ = 40 \text{ yd}^2$

The cost of 40 yd² of carpeting is $30(40) = $1,200$.

57. The area of tile needed is the area of the rectangular room.

$A = lw$

$A = (14 \text{ ft})(20 \text{ ft})$

$ = 280 \text{ ft}^2$

The cost of 280 ft² of floor tiles is $1.29(280) = 361.20.

59. Convert 12 feet and 24 feet to yards.

$$12 \text{ ft} = 12 \text{ ft} \cdot \frac{1 \text{ yd}}{3 \text{ ft}}$$

$$= \frac{12}{3} \text{ yd}$$

$$= 4 \text{ yd}$$

$$24 \text{ ft} = 24 \text{ ft} \cdot \frac{1 \text{ yd}}{3 \text{ ft}}$$

$$= \frac{24}{3} \text{ yd}$$

$$= 8 \text{ yd}$$

The area of the triangular sail is given by the formula $A = \frac{1}{2} bh$.

$A = \frac{1}{2} bh$

$A = \frac{1}{2}(12 \text{ ft})(24 \text{ ft})$

$ = \frac{1}{2}(4 \text{ yd})(8 \text{ yd})$

$ = 16 \text{ yd}^2$

The cost of 16 yd² of nylon is $16 \cdot \$12 = \$192$.

61. Convert 15 yards, 9 yards, and 20 yards to feet.

$$15 \text{ yd} = 15 \text{ yd} \cdot \frac{3 \text{ ft}}{1 \text{ yd}}$$

$$= 15 \cdot 3 \text{ ft}$$

$$= 45 \text{ ft}$$

$$9 \text{ yd} = 9 \text{ yd} \cdot \frac{3 \text{ ft}}{1 \text{ yd}}$$

$$= 9 \cdot 3 \text{ ft}$$

$$= 27 \text{ ft}$$

$$20 \text{ yd} = 20 \text{ yd} \cdot \frac{3 \text{ ft}}{1 \text{ yd}}$$

$$= 20 \cdot 3 \text{ ft}$$

$$= 60 \text{ ft}$$

The area of a trapezoid is given by the formula $A = \frac{1}{2}h(b_1 + b_2)$.

$$A = \frac{1}{2}h(b_1 + b_2)$$

$$A = \frac{1}{2}(9 \text{ yd})(20 \text{ yd} + 15 \text{ yd})$$

$$= \frac{1}{2}(27 \text{ ft})(60 \text{ ft} + 45 \text{ ft})$$

$$= \frac{1}{2}(27 \text{ ft})(105 \text{ ft})$$

$$= \frac{2,835}{2} \text{ ft}^2$$

$$= 1,417.5 \text{ ft}^2$$

The cost of 1,417.5 ft^2 of sod is 1,417.5 • \$1.17 = \$1,658.475. It will cost \$1,658.48 to sod the lawn.

63. The front (containing the doors) and back of the first floor are rectangles with length 20 ft and height 12 ft.

$$A_{\text{front}} = lw$$

$$A_{\text{front}} = (20)(12)$$

$$= 240$$

The front and back walls each have area 240 ft². The side walls are rectangles with length 48 ft and height 12 ft.

$$A_{\text{side}} = l \cdot w$$

$$A_{\text{side}} = (48)(12)$$

$$= 576$$

The side walls each have area 576 ft². The front, back, and two sides are to be covered.

$$A = A_{\text{front}} + A_{\text{back}} + A_{\text{side 1}} + A_{\text{side 2}}$$
$$= 240 + 240 + 576 + 576$$
$$= 1{,}632$$

The area to be covered with sheetrock is 1,632 ft². Each sheet of sheetrock covers 4 ft • 8 ft = 32 ft², so 1 sheet is needed per 32 ft² of wall to be covered.

$$1{,}632 \text{ ft}^2 \cdot \frac{1 \text{ sheet}}{32 \text{ ft}^2} = \frac{1{,}632}{32} \text{ sheets}$$

$$= 51 \text{ sheets}$$

51 sheets of sheetrock will be needed.

REVIEW

67.
$$\frac{3}{4} + \frac{2}{3} = \frac{3 \cdot 3}{4 \cdot 3} = \frac{2 \cdot 4}{3 \cdot 4}$$

$$= \frac{9}{12} + \frac{8}{12}$$

$$= \frac{9 + 8}{12}$$

$$= \frac{17}{12}$$

$$= 1\frac{5}{12}$$

69.
$$3\frac{3}{4} + 2\frac{1}{3} = \frac{15}{4} + \frac{7}{3}$$

$$= \frac{15 \cdot 3}{4 \cdot 3} + \frac{7 \cdot 4}{3 \cdot 4}$$

$$= \frac{45}{12} + \frac{28}{12}$$

$$= \frac{45 + 28}{12}$$

$$= \frac{73}{12}$$

$$= 6\frac{1}{12}$$

71.
$$7\frac{1}{2} \div 5\frac{2}{5} = \frac{15}{2} \div \frac{27}{5}$$
$$= \frac{15}{2} \cdot \frac{5}{27}$$
$$= \frac{15 \cdot 5}{2 \cdot 27}$$
$$= \frac{3 \cdot 5 \cdot 5}{2 \cdot 3 \cdot 9}$$
$$= \frac{5 \cdot 5}{2 \cdot 9}$$
$$= \frac{25}{18}$$
$$= 1\frac{7}{18}$$

STUDY SET Section 9.6
VOCABULARY

1. A segment drawn from the <u>center</u> of a circle to a point on the circle is called a <u>radius</u>.

3. A <u>diameter</u> is a chord that passes hrough the <u>center</u> of a circle.

5. An arc that is shorter than a <u>semicircle</u> is called a <u>minor</u> arc.

7. The distance around a circle is called its <u>circumference</u>.

CONCEPTS

9. Since the center of the circle is O, OA, OB, and OC are radii of the circle.

11. A, C, and D are points on the circle, so DA, DC, and AC are chords.

13. The semicircles are $\overset{\frown}{ABC}$ and $\overset{\frown}{ADC}$.

15. The diameter of a circle can be found by doubling the radius.

NOTATION

17. The symbol $\overset{\frown}{AB}$ is read as "arc AB."

19. The formula for the circumference of a circle is $\underline{C = \pi D}$ or $\underline{C = 2\pi r}$.

PRACTICE

21. $C = \pi D$
$C = \pi(12)$
$\quad = 12\pi$
$\quad \approx 37.70$
The circumference is approximately 37.70 in.

23. $\quad C = \pi D$
$36\pi = \pi D$
$\quad 36 = D$
The diameter is 36 meters.

25. The figure is a combination of two sides of a rectangle and two semicircles. Both semicircles have diameter of 3 ft, so the two semicircles combine to form a whole circle of diameter 3 ft. The rectangle contributes two sides of length 8 ft.

$$P_{\text{rectangular part}} = 8 + 8$$
$$= 16$$
$$P_{\text{circle}} = \pi D$$
$$= \pi(3)$$
$$= 3\pi$$
$$\approx 9.42$$

The perimeter of the figure is
$$P = P_{\text{rectangle}} + P_{\text{circle}}$$
$$\approx 16 + 9.42$$
$$\approx 25.42$$

The perimeter is about 25.42 cm.

27. The figure is a combination of hree sides of a rectangle and a semicircle.

$$P_{\text{rectangular part}} = 8 + 6 + 8$$
$$= 22$$

The diameter of the semicircle is 6 meters.

$$P_{\text{semicircle}} = \frac{1}{2}\pi D$$

$$= \frac{1}{2}\pi(6)$$

$$= 3\pi$$

The perimeter of the figure is
$$P = P_{\text{rectangular part}} + P_{\text{semicircle}}$$
$$= 22 + 3\pi$$
$$\approx 31.42$$

The perimeter is about 31.42 m.

29. The area of a circle is given by $A = \pi r^2$.

$$A = \pi r^2$$
$$A = \pi(3)^2$$
$$= \pi(9)$$
$$= 9\pi$$
$$\approx 28.3$$

The area is about 28.3 in.^{2}.

31. The figure is a combination of a rectangle and two semicircles. The rectangle has length 10 in. and width 6 in.

$$A_{\text{rectangle}} = lw$$
$$= (10)(6)$$
$$= 60$$

The semicircles both have diameter 6 in., so they combine to form a complete circle with diameter 6 in. The radius of the circle is

$$\frac{1}{2}(6 \text{ in.}) = 3 \text{ in.}$$

$$A_{\text{circle}} = \pi r^2$$
$$= \pi(3)^2$$
$$= \pi(9)$$
$$= 9\pi$$
$$\approx 28.3$$

The total area of the figure is
$$A = A_{\text{rectangle}} + A_{\text{circle}}$$
$$\approx 60 + 28.3$$
$$\approx 88.3$$

The area is about 88.3 in.^{2}.

33. The figure is a combination of a triangle and a semicircle. The triangle has base of length 12 cm and height 12 cm.

$$A_{\text{triangle}} = \frac{1}{2}bh$$

$$= \frac{1}{2}(12)(12)$$

$$= 72$$

The semicircle has diameter 12 cm, so the radius is

$$\frac{1}{2}(12 \text{ cm}) = 6 \text{ cm}.$$

$$A_{\text{semicircle}} = \frac{1}{2}\pi r^2$$

$$= \frac{1}{2}\pi(6)^2$$

$$= \frac{1}{2}\pi(36)$$

$$= 18\pi$$

$$\approx 56.5$$

The total area of the figure is

$$A = A_{\text{triangle}} + A_{\text{semicircle}}$$

$$\approx 72 + 56.5$$

$$\approx 128.5$$

The area is about 128.5 cm².

35. The shaded region is a rectangle with a circle taken out. The rectangle has length 4 in. and width 10 in.

$$A_{\text{rectangle}} = lw$$

$$= (4)(10)$$

$$= 40$$

The circle has diameter 4 in., so the radius is $\frac{1}{2}(4 \text{ in.}) = 2 \text{ in.}$

$$A_{\text{circle}} = \pi r^2$$

$$= \pi(2)^2$$

$$= \pi(4)$$

$$= 4\pi$$

The area of the shaded region is

$$A = A_{\text{rectangle}} - A_{\text{circle}}$$

$$= 40 - 4\pi$$

$$\approx 27.4$$

The area is about 27.4 in.².

37. The shaded region is a parallelogram with a circle taken out. The parallelogram has base length 13 in. and height 9 in.

$$A_{\text{parallelogram}} = bh$$

$$= (13)(9)$$

$$= 117$$

The circle has radius 4 in.

$$A_{\text{circle}} = \pi r^2$$

$$= \pi(4)^2$$

$$= \pi(16)$$

$$= 16\pi$$

The area of the shaded region is

$$A = A_{\text{parallelogram}} - A_{\text{circle}}$$

$$= 117 - 16\pi$$

$$\approx 66.7$$

The area is about 66.7 in.².

APPLICATIONS

39. The diameter of Round Lake is 2 miles, so the radius is

$$\frac{1}{2}(2 \text{ mi}) = 1 \text{ mi.}$$

$$A = \pi r^2$$
$$A = \pi (1)^2$$
$$= \pi (1)$$
$$= \pi$$
$$\approx 3.14$$

The area of Round Lake is about 3.14 mi².

41. The distance around the track is its circumference.

$$C = \pi D$$

$$C = \pi\left(\frac{1}{4}\right)$$

$$= \frac{1}{4}\pi$$

Each time Joan jogs around the track, she goes $\frac{1}{4}\pi$ miles, so there are $\frac{1}{4}\pi$ miles per lap around the track.

$$10 \text{ miles} \cdot \frac{1 \text{ lap}}{\frac{1}{4}\pi \text{ miles}} = \frac{10}{\frac{1}{4}\pi} \text{ laps}$$

$$= 10 \div \frac{1}{4}\pi \text{ laps}$$

$$= 10 \div \frac{\pi}{4} \text{ laps}$$

$$= 10 \cdot \frac{4}{\pi} \text{ laps}$$

$$= \frac{40}{\pi} \text{ laps}$$

$$\approx 12.73 \text{ laps}$$

Joan must circle the track about 12.73 times.

43. Let r_1 = the radius of the earth at the equator. Then the circumference of the earth at the equator is $C = 2\pi r_1$. This is also the length of the original steel band. The looser steel band will have length $C + 10$. Let r_2 = the radius of the loose steel band. We want to know the difference between the radii of the two bands, $r_2 - r_1$. Since $C + 10$ is the circumference of the band with radius r_2, $C + 10 = 2\pi r_2$.

Solve $C = 2\pi r_1$ for r_1:

$$2\pi r_1 = C$$

$$r_1 = \frac{C}{2\pi}$$

Solve $C + 10 = 2\pi r_2$ for r_2:

$$2\pi r_2 = C + 10$$

$$r_2 = \frac{C + 10}{2\pi}$$

$$r_2 - r_1 = \frac{C + 10}{2\pi} - \frac{C}{2\pi}$$

$$= \frac{C + 10 - C}{2\pi}$$

$$= \frac{10}{2\pi}$$

$$\approx 1.59$$

The band will be about 1.59 feet above the earth's surface.

45. The bullseye has diameter 1 foot, so the radius is $\frac{1}{2}(1 \text{ ft}) = \frac{1}{2}$ ft. The area of the bullseye is

$$A_{\text{bullseye}} = \pi r^2$$

$$= \pi\left(\frac{1}{2}\right)^2$$

$$= \pi\left(\frac{1}{4}\right)$$

$$= \frac{1}{4}\pi$$

The target has diameter 4 feet, so the radius is $\frac{1}{2}(4 \text{ ft}) = 2$ ft. The area of the target is

$$A_{\text{target}} = \pi r^2$$

$$= \pi(2)^2$$

$$= \pi(4)$$

$$= 4\pi$$

The percentage is the ratio of the area of the bullseye to the area of the target.

$$\frac{A_{\text{bullseye}}}{A_{\text{target}}} = \frac{\frac{1}{4}\pi}{4\pi}$$

$$= \frac{1}{4} \div 4$$

$$= \frac{1}{4} \cdot \frac{1}{4}$$

$$= \frac{1}{16}$$

$$= 0.0625$$

$$= 6.25\%$$

The bullseye is 6.25% of the area of the target.

REVIEW

51.

$$\begin{array}{r} 0.9 \\ 10\overline{)9.0} \\ \underline{9\,0} \\ 0 \end{array}$$

$$\frac{9}{10} = 0.9$$
$$= 90\%$$

53. 30% of 1,600 is what number?

$$30\% \cdot 1,600 = x$$
$$0.30 \cdot 1,600 = x$$
$$480 = x$$

30% of 1,600 is 480.

55. 25% of $98 is what number?

$$25\% \cdot 98 = x$$
$$0.25 \cdot 98 = x$$
$$24.5 = x$$

The discount is $24.50.
Maria paid $98 − $24.50 = $73.50.

STUDY SET Section 9.7

VOCABULARY

1. The space contained within a geometric solid is called its <u>volume</u>.

3. A <u>cube</u> is a rectangular solid with all sides of equal length.

5. The <u>surface area</u> of a rectangular solid is the <u>sum</u> of the areas of its faces.

7. A <u>cylinder</u> is a hollow figure like a drinking straw.

9. A <u>cone</u> looks like a witch's pointed hat.

CONCEPTS

11. $V = lwh$

13. $V = \dfrac{4}{3}\pi r^3$

15. $V = \dfrac{1}{3}Bh$

17. $SA = 2lw + 2lh + 2hw$

19. 1 cubic yard is the volume of a cube that is 1 yard on each side. 1 yard = 3 feet, so 1 cubic yard is the volume of a cube that is 3 feet on each side.

$$V = s^3$$
$$V = (3)^3$$
$$= 27$$
$$1\ \text{yd}^3 = 27\ \text{ft}^3$$

21. 1 cubic meter is the volume of a cube that is 1 meter on each side. 1 meter = 10 decimeters 1 cubic meter is the volume of a cube that is 10 decimeters on each side.

$$V = s^3$$
$$V = (10)^3$$
$$= 1,000$$
$$1\ \text{m}^3 = 1,000\ \text{dm}^3$$

23. a. Volume

    b. Area

    c. Volume

    d. Surface Area

    e. Perimeter

    f. Surface Area

NOTATION

25. The symbol in.^{3} is read as "<u>1 cubic inch</u>."

PRACTICE

27. $V = lwh$
$V = 3 \cdot 4 \cdot 5$
$= 60$
The volume is 60 cm^3.

29. $V = Bh$
B is the area of a triangle with base 3 meters and height 4 meters.

$B = \dfrac{1}{2}bh$

$B = \dfrac{1}{2}(3)(4)$

$= 6$
$V = Bh$
$V = (6)(8)$
$= 48$
The volume is 48 m^3.

31. $V = \dfrac{4}{3}\pi r^3$

$V = \dfrac{4}{3}\pi(9)^3$

$= \dfrac{4}{3}\pi(729)$

$= 972\pi$

$\approx 3,053.63$
The volume is about 3,053.63 in.^{3}.

33. $V = Bh$
B is the area of a circle with radius 6 meters.
$B = \pi r^2$
$B = \pi(6)^2$
$= \pi(36)$
$= 36\pi$
$V = Bh$
$V = (36\pi)(12)$
$= 432\pi$
$\approx 1,357.17$
The volume is about 1,357.17 m^3.

35. $V = \dfrac{1}{3}Bh$

B is the area of a circle with diameter 10 centimeters. The radius of the base is

$\dfrac{1}{2}(10 \text{ cm}) = 5 \text{ cm}$.

$B = \pi r^2$
$B = \pi(5)^2$
$= \pi(25)$
$= 25\pi$

$V = \dfrac{1}{3}Bh$

$V = \dfrac{1}{3}(25\pi)(12)$

$= 100\pi$

≈ 314.16
The volume is about 314.16 cm^3.

37. $V = \dfrac{1}{3}Bh$

B is the area of a square with sides of length 10 meters.

$B = s^2$

$B = (10)^2$
$\quad = 100$

$V = \dfrac{1}{3}Bh$

$V = \dfrac{1}{3}(100)(12)$

$\quad = 400$

The volume is 400 m^3.

39. $SA = 2lw + 2lh + 2wh$
$SA = 2 \cdot 3 \cdot 4 + 2 \cdot 3 \cdot 5 + 2 \cdot 4 \cdot 5$
$\quad = 24 + 30 + 40$
$\quad = 94$

The surface area is 94 cm^2.

41. $SA = 4\pi r^2$

$SA = 4\pi(10)^2$
$\quad = 4\pi(100)$
$\quad = 400\pi$
$\quad \approx 1,256.64$

The surface area is about 1,256.64 in.^{2}.

43. The figure is a cube combined with a pyramid with a square base. The cube has sides of length 8 centimeters.

$V_{\text{cube}} = s^3$
$\quad\quad = (8)^3$
$\quad\quad = 512$

The base of the pyramid is a square with sides of length 8 cm.

$B = s^2$

$B = (8)^2$
$\quad = 64$

The pyramid has a base with area 64 cm^2 and height 3 cm.

$V_{\text{pyramid}} = \dfrac{1}{3}Bh$

$\quad\quad = \dfrac{1}{3}(64)(3)$

$\quad\quad = 64$

$V = V_{\text{cube}} + V_{\text{pyramid}}$
$\quad = 512 + 64$
$\quad = 576$

The volume is 576 cm^3.

45. The figure is a combination of two identical cones. The base of each cone is a circle with diameter 8 in.

The radius is $\frac{1}{2}(8 \text{ in.}) = 4$ in.

$B = \pi r^2$
$B = \pi(4)^2$
$\quad = \pi(16)$
$\quad = 16\pi$

Each cone has a base with area 16π in.^{2} and height 10 in.

$V_{\text{cone}} = \frac{1}{3}Bh$

$\quad\quad\ = \frac{1}{3}(16\pi)(10)$

$\quad\quad\ = \frac{160}{3}\pi$

The volume of the figure is double the volume of one cone.

$V = 2 \cdot V_{\text{cone}}$

$\quad = 2 \cdot \frac{160}{3}\pi$

$\quad = \frac{320}{3}\pi$

$\quad \approx 335.10$

The volume is about 335.10 in.^{3}.

APPLICATIONS

47. $V = s^3$

$V = \left(\frac{1}{2}\right)^3$

$\quad = \frac{1}{8}$

$\quad = 0.125$

The volume is 0.125 in.^{3}.

49. $V = Bh$

The base of the cylinder is a circle with diameter 6 feet. The radius is $\frac{1}{2}(6 \text{ ft}) = 3$ ft.

$B = \pi r^2$
$B = \pi(3)^2$
$\quad = \pi(9)$
$\quad = 9\pi$

In this case the cylinder has a base with area 9π ft^2 and height 7 ft.

$V = Bh$
$V = (9\pi)(7)$
$\quad = 63\pi$
$\quad \approx 197.92$

The volume is about 197.92 ft^3.

51. $V = \frac{4}{3}\pi r^3$

The diameter of the balloon is 40 feet, so the radius is

$\frac{1}{2}(40 \text{ ft}) = 20$ ft.

$V = \frac{4}{3}\pi r^3$

$V = \frac{4}{3}\pi(20)^3$

$\quad = \frac{4}{3}\pi(8,000)$

$\quad = \frac{32,000}{3}\pi$

$\quad \approx 33,510.32$

The balloon will hold about 33,510.32 ft^3 of gas.

REVIEW

55. $4(6+4) - 2^2 = 4(6+4) - 4$
$\quad\quad\quad\quad\quad\quad\ = 4(10) - 4$
$\quad\quad\quad\quad\quad\quad\ = 40 - 4$
$\quad\quad\quad\quad\quad\quad\ = 36$

57. $5+2(6+2^3) = 5+2(6+8)$
$$ = 5+2(14)$$
$$ = 5+28$$
$$ = 33$$

59. The six pencils and the notebook cost
$$6 \cdot \$0.60 + \$1.25 = \$3.60 + \$1.25$$
$$ = \$4.85$$
The charge that Carlos received was
$\$5.00 - \$4.85 = \$0.15$.

61. The 3 packages of golf balls, 1 package of tees, and the golf glove cost
$$3 \cdot \$1.99 + \$0.49 + \$6.95 = \$5.97 + \$0.49 + \$6.95$$
$$ = \$6.46 + \$6.95$$
$$ = \$13.41$$
George spent $13.41.

KEY CONCEPT Formulas

1. $d = rt$

3. $P = 2l + 2w$

5. The base of the triangular plot has length 700 ft, and the height is 600 ft.

$$A = \frac{1}{2}bh$$

$$A = \frac{1}{2}(700)(600)$$

$$ = 210,000$$

The area is 210,000 ft^3.

7. The retail price is the store owner's cost plus the mark-up.

$$p = c + m$$
$$p = \$45.50 + \$35.00$$
$$ = \$80.50$$

The retail price is $80.50.

9. Distance fallen is 16 times (time)^2.

$$d = 16t^2$$
$$d = 16(3)^2$$
$$ = 16(9)$$
$$ = 144$$

The rock falls 144 feet.

11. $I = Prt$

For the savings account, $P = 5,000$, $r = 5\% = 0.05$, and $t = 3$.

$I = Prt$
$= 5,000 \cdot 0.05 \cdot 3$
$= 750$

The interest earned is $750.

For the passbook account, $P = 2,250$, $r = 2\% = 0.02$, and $t = 1$.

$I = Prt$
$= 2,250 \cdot 0.02 \cdot 1$
$= 45$

The interest earned is $45.

For the trust fund, $P = 10,000$, $r = 6.25\% = 0.0625$, and $t = 10$.

$I = Prt$
$= 10,000 \cdot 0.0625 \cdot 10$
$= 6,250$

The interest earned is $6,250.

Type of account	Principal	Annual rate earned	Time invested	Interest earned
Savings	$5,000	5%	3 yr	$750
Passbook	$2,250	2%	1 yr	$45
Trust Fund	$10,000	6.25%	10 yr	$6,250

CHAPTER 9 REVIEW

1. C and D are points, CD is a line, and AB is a plane.

3. The angle can be named $\angle ABC$, $\angle CBA$, $\angle B$, and $\angle 1$.

5. $\angle 1$ and $\angle 2$ are acute, $\angle ABD$ and $\angle CBD$ are right angles, $\angle CBE$ is an obtuse angle, and $\angle ABC$ is a straight angle.

7. The sum of the 35° angle and the $x°$ angle is 50°.
$35 + x = 50$
$x = 15$

9. a. $m(\angle 1) = 65°$ since $\angle 1$ and the 65° angle are vertical angles.

b. $m(\angle 2) = 115°$ since $\angle 2$ and the 65° angle are supplementary angles.

11. Let $x° =$ the measure of the supplementary angle.
$x + 140 = 180$
$x = 40$
The supplement of a 140° angle is a 40° angle.

13. Part (a) represents parallel lines.

15. The pairs of corresponding angles are $\angle 1$ and $\angle 5$, $\angle 2$ and $\angle 6$, $\angle 3$ and $\angle 7$, $\angle 4$ and $\angle 8$.

17. $m(\angle 1) = 70°$ since $\angle 1$ and the $110°$ angle are supplementary.
$m(\angle 2) = 110°$ since $\angle 2$ and the $110°$ angle are vertical angles.
$m(\angle 3) = 70°$ since $\angle 3$ and the $110°$ angle are supplementary.
$m(\angle 4) = 110°$ since $\angle 4$ and the $110°$ angle are alternate interior angles formed by a transversal of parallel lines.
$m(\angle 5) = 70°$ since $\angle 5$ and $\angle 4$ are supplementary.
$m(\angle 6) = 110°$ since $\angle 6$ and $\angle 4$ are vertical angles.
$m(\angle 7) = 70°$ since $\angle 7$ and $\angle 4$ are supplementary.

19. The $(x + 10)°$ angle and the $(2x - 30)°$ angle are alternate interior angles.
$$x + 10 = 2x - 30$$
$$10 = x - 30$$
$$40 = x$$

21. a. The polygon has 8 sides so it is an octagon.

b. The polygon has 5 sides so it is a pentagon.

c. The polygon has 3 sides so it is a triangle.

d. The polygon has 6 sides so it is a hexagon.

e. The polygon has 4 sides so it is a quadrilateral.

23. a. Two sides have equal length, so the triangle is an isosceles triangle.

b. The three sides have different lengths, so the triangle is a scalene triangle.

c. All three sides have equal length, so the triangle is an equilateral triangle.

d. One angle is a right angle, so the triangle is a right triangle.

25. a.
$$70 + 20 + x = 180$$
$$90 + x = 180$$
$$x = 90$$

b.
$$70 + 60 + x = 180$$
$$130 + x = 180$$
$$x = 50$$

27. The base angles of an isosceles triangle have equal measure, so if $x° =$ the measure of the vertex angle, the angles of the triangle have measures $60°$, $60°$, and x.
$$60 + 60 + x = 180$$
$$120 + x = 180$$
$$x = 60$$
All three angles of the triangle have measure $60°$, so the triangle is an equilateral triangle.

29. a. $m\left(\overline{BD}\right) = 15$ cm since the diagonals of a rectangle have equal length.

b. $m(\angle 1) = 40°$ since $\angle 1$ and the $40°$ angle are congruent alternate interior angles.

29. c. $\angle ECD$ and $\angle ECB$ are complementary.

$$m(\angle ECD) + m(\angle ECB) = 90°$$
$$m(\angle ECD) + 50° = 90°$$
$$m(\angle ECD) = 40°$$

$\angle EDC$, $\angle DEC$, and $\angle ECD$ are the angles of $\triangle EDC$.

$$m(\angle EDC) + m(\angle DEC) + m(\angle ECD) = 180°$$
$$40° + m(\angle DEC) + 40° = 180°$$
$$80° + m(\angle DEC) = 180°$$
$$m(\angle DEC) = 100°$$

$\angle 2$ and $\angle DEC$ are vertical angles so $m(\angle 2) = 100°$.

31. a. $m(\angle B) = 65°$ since the trapezoid is an isosceles trapezoid.

   b. $m(\angle C) = 115°$ since $\angle B$ and $\angle C$ are supplementary because they are interior angles on the same side of a transversal of parallel lines.

33. $\angle A$ corresponds to $\angle D$.

$\angle B$ corresponds to $\angle E$.

$\angle C$ corresponds to $\angle F$.

$\overline{AC}$ corresponds to $\overline{DF}$.

$\overline{AB}$ corresponds to $\overline{DE}$.

$\overline{BC}$ corresponds to $\overline{EF}$.

35. a. The triangles are similar since two pairs of angles are congruent.

   b. The triangles are similar since two pairs of angles are congruent. One pair of congruent angles are the vertical angles at the vertex where the triangles meet.

37. a.
$$c^2 = a^2 + b^2$$
$$c^2 = (5)^2 + (12)^2$$
$$c^2 = 25 + 144$$
$$c^2 = 169$$
$$\sqrt{c^2} = \sqrt{169}$$
$$c = 13$$

   b.
$$a^2 + b^2 = c^2$$
$$(8)^2 + b^2 = (17)^2$$
$$64 + b^2 = 289$$
$$b^2 = 225$$
$$\sqrt{b^2} = \sqrt{225}$$
$$b = 15$$

39. For a square, $P = 4s$.
$$P = 4(18)$$
$$= 72$$
The perimeter is 72 in.

41. a. $P = 8 + 4 + 4 + 8 + 6$
$$= 30$$
The perimeter is 30 m.

41. b. The figure is a rectangle (on the left) combined with a square. The unmarked top has length 10 m and the unmarked vertical segment has length 4 m.
$$P = 10 + 4 + 4 + 4 + 6 + 8$$
$$= 36$$
The perimeter is 36 m.

43. One square yard is the area of a square that is one yard on each side. 1 yard is 3 feet, so one square yard is the area of a square that is
3 feet on each side.
$$A = s^2$$
$$A = (3)^2$$
$$= 9$$
One square yard is 9 square feet.

45. a. $\overline{AB}$ and $\overline{CD}$ are chords.

b. $\overline{AB}$ is a diameter.

c. $\overline{OA}, \overline{OC}, \overline{OD},$ and $\overline{OB}$ are radii.

d. The center of the circle is O.

47. The figure is a rectangle with a semicircle on each end.
The two sides of the rectangle that contribute to the perimeter each have length 10 cm.
$$P_{\text{rectangular part}} = 10 + 10$$
$$= 20$$
The two semicircles combine to form a complete circle with diameter 8 cm.
$$P_{\text{circle}} = \pi D$$
$$= \pi(8)$$
$$= 8\pi$$
$$P = P_{\text{rectangular part}} + P_{\text{circle}}$$
$$= 20 + 8\pi$$
$$\approx 45.1$$
The perimeter is about 45.1 cm.

49. The figure is a combination of a rectangle and two semicircles. The rectangle has length 10 cm and length 8 cm.
$$A_{\text{rectangle}} = lw$$
$$= (10)(8)$$
$$= 80$$
The semicircles combine to form a complete circle with diameter 8 cm. The radius is
$$\frac{1}{2}(8 \text{ cm}) = 4 \text{ cm}.$$
$$A_{\text{circle}} = \pi r^2$$
$$= \pi(4)^2$$
$$= \pi(16)$$
$$= 16\pi$$
$$A = A_{\text{rectangle}} + A_{\text{circle}}$$
$$= 80 + 16\pi$$
$$\approx 130.3$$
The area is about 130.3 cm^2.

51. One cubic foot is the volume of a cube that is 1 foot on each side. One foot is 12 inches, so one cubic foot is the volume of a cube that is 12 inches on each side.
$$V = s^3$$
$$V = (12)^3$$
$$= 1,728$$
One cubic foot is 1,728 cubic inches.

53. a. For a rectangular solid, $SA = 2lw + 2lh + 2wh$.

$$SA = 2(3.1)(2.3) + 2(3.1)(4.4) + 2(2.3)(4.4)$$
$$= 14.26 + 27.28 + 20.24$$
$$= 61.78$$
$$\approx 61.8$$

The surface area is about 61.8 ft^2.

b. For a sphere, $SA = 4\pi r^2$.

$$SA = 4\pi(5)^2$$
$$= 4\pi(25)$$
$$= 100\pi$$
$$\approx 314.2$$

The surface area is about 314.2 in.^{2}.

CHAPTER 9 TEST

1. $m(AB) = 7 - 3$
$$= 4$$

3. True; the measure of an acute angle is between 0° and 90°.

5. False; the measure of a right angle is 90°.

7. The sum of the $x°$ angle and the 17° angle is 67°.
$$x + 17 = 67$$
$$x = 50$$

9. Since they are vertical angles, the $(3y + 4)°$ angle and the $(5y - 20)°$ angle have equal measures.
$$3y + 4 = 5y - 20$$
$$4 = 2y - 20$$
$$24 = 2y$$
$$12 = y$$

11. The measures of complementary angles sum to 90°.
Let $x° =$ the measure of the angle complementary to a 67° angle.
$$67 + x = 90$$
$$x = 23$$
The complement of an angle measuring 67° is an angle measuring 23°.

13. $m(\angle 1) = 70°$ since $\angle 1$ and the 70° angle are alternate interior angles formed by a transversal of parallel lines.

15. $m(\angle 3) = 70°$ since $\angle 3$ and the 70° angle are corresponding angles formed by a transversal of parallel lines.

17.

Polygon	Number of sides
Triangle	3
Quadrilateral	4
Hexagon	6
Pentagon	5
Octagon	8

19. $m(\angle A) = 57°$ since the triangle is isosceles.

21. The sum of the measures of the angles of a triangle is 180°. Let $x°$ = the measure of the third angle of the triangle.
$$65° + 85° + x° = 180°$$
$$150° + x° = 180°$$
$$x° = 30°$$
The third angle has measure 30°.

23. Opposite sides have equal length and the diagonals have equal length.
$$m(\overline{AB}) = m(\overline{DC}),$$
$$m(\overline{AD}) = m(\overline{BC}), \text{ and}$$
$$m(\overline{AC}) = m(\overline{BD})$$

25. $m(\overline{DE}) = 8$ in. since $\overline{DE}$ corresponds to $\overline{AB}$.

27. The triangles are similar, so corresponding sides are in proportion.
$$\frac{m(\overline{DE})}{m(\overline{AB})} = \frac{m(\overline{DF})}{m(\overline{AC})}$$
$$\frac{x}{9} = \frac{4}{6}$$
$$x \cdot 6 = 9 \cdot 4$$
$$6x = 36$$
$$x = 6$$

29. The distance from first base to third base is the length of the hypotenuse of a right triangle whose legs both have length 90 ft. Use the Pythagorean theorem with $a = 90$ and $b = 90$.
$$c^2 = a^2 + b^2$$
$$c^2 = (90)^2 + (90)^2$$
$$c^2 = 8,100 + 8,100$$
$$c^2 = 16,200$$
$$\sqrt{c^2} = \sqrt{16,200}$$
$$c \approx 127.3$$
The distance from first base to third base is about 127.3 ft.

31. For a trapezoid, $A = \frac{1}{2}h(b_1 + b_2)$.
$$A = \frac{1}{2}(6)(12.2 + 15.7)$$
$$= \frac{1}{2}(6)(27.9)$$
$$= 83.7$$
The area is 83.7 ft^2.

33. For a circle, $A = \pi r^2$. The diameter of the circle is 6 feet, so the radius is $\frac{1}{2}(6 \text{ ft}) = 3$ ft.

$$A = \pi(3)^2$$
$$= \pi(9)$$
$$= 9\pi$$
$$\approx 28.3$$

The area is about 28.3 ft^2.

35. For a sphere, $V = \frac{4}{3}\pi r^3$. The diameter of the sphere is 8 meters, so the radius is $\frac{1}{2}(8 \text{ m}) = 4$ m.

$$V = \frac{4}{3}\pi(4)^3$$

$$= \frac{4}{3}\pi(64)$$

$$= \frac{256}{3}\pi$$

$$\approx 268.1$$

The volume is about 268.1 m^3.

CHAPTERS 1–9
CUMULATIVE REVIEW

1. $x + y + z = 2 + (-3) - (-4)$
$$= 2 - 3 + 4$$
$$= -1 + 4$$
$$= 3$$

3.
$$3(x + 2) = 13$$
$$3x + 3 \cdot 2 = 13$$
$$3x + 6 = 13$$
$$3x + 6 - 6 = 13 - 6$$
$$3x = 7$$
$$\frac{3x}{3} = \frac{7}{3}$$
$$x = \frac{7}{3}$$

5.
$$3(p + 15) + 4(11 - p) = 0$$
$$3p + 3 \cdot 15 + 4 \cdot 11 - 4p = 0$$
$$3p + 45 + 44 - 4p = 0$$
$$(3p - 4p) + (45 + 44) = 0$$
$$-p + 89 = 0$$
$$-p + 89 + p = 0 + p$$
$$89 = p$$

7. $x^4 x^5 = x^{4+5}$
$$= x^9$$

9. $(p^2)^3 (p^3)^2 = p^{2 \cdot 3} p^{3 \cdot 2}$
$$= p^6 p^6$$
$$= p^{6+6}$$
$$= p^{12}$$

11. $3a + 4b - 5a + 2b = (3a - 5a) + (4b + 2b)$
$$= -2a + 6b$$

13. $\dfrac{2x}{3} \cdot \dfrac{y}{5} = \dfrac{2x \cdot y}{3 \cdot 5}$

$\qquad = \dfrac{2xy}{15}$

15. $\dfrac{x}{4} - \dfrac{3}{5} = \dfrac{x \cdot 5}{4 \cdot 5} - \dfrac{3 \cdot 4}{5 \cdot 4}$

$\qquad = \dfrac{5x}{20} - \dfrac{12}{20}$

$\qquad = \dfrac{5x - 12}{20}$

17.

$$
\begin{array}{r}
0.4 \\
15\overline{)6.0} \\
\underline{6\,0} \\
0
\end{array}
$$

$\dfrac{6}{15} = 0.4$

19.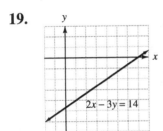

21. $(5x^2 - 2x + 4) - (3x^2 - 5) = (5x^2 - 2x + 4) + [-(3x^2 - 5)]$

$\qquad\qquad\qquad\qquad\qquad\quad = (5x^2 - 2x + 4) + (-3x^2 + 5)$

$\qquad\qquad\qquad\qquad\qquad\quad = (5x^2 - 3x^2) + (-2x) + (4 + 5)$

$\qquad\qquad\qquad\qquad\qquad\quad = 2x^2 - 2x + 9$

23. $-3p(2p^2 + 3p - 4) = -3p(2p^2) - 3p(3p) - 3p(-4)$

$\qquad\qquad\qquad\qquad\;\; = -6p^3 - 9p^2 + 12p$

25. What number is 37% of 460?

$x = 37\% \cdot 460$

$x = 0.37 \cdot 460$

$x = 170.2$

37% of 460 is 170.2.

27. $150 = x \cdot 600$

$\dfrac{150}{600} = x$

$0.25 = x$

$25\% = x$

150 is 25% of 600.

29. What number is $6\dfrac{1}{4}\%$ of $18,550?

$x = 6\dfrac{1}{4}\% \cdot 18,550$

$x = 6.25\% \cdot 18,550$

$x = 0.0625 \cdot 18,550$

$x = 1,159.375$

$1,159.38 in sales tax will be added to the price of the car.

31. $10,124.22 is what percent of $47,000?

$10,124.22 = x \cdot 47,000$

$\dfrac{10,124.22}{47,000} = x$

$0.2154 \approx x$

$21.5\% \approx x$

The man's tax rate is about 21.5%.

33. For the discount, what number is 30% of $69.90?

$x = 30\% \cdot 69.90$

$x = 0.30 \cdot 69.90$

$x = 20.97$

The discount is $20.97. The sale price is

$69.90 − $20.97 = $48.93.

35. Use the formula $I = Prt$ with $P = 1,500$, $r = 9\% = 0.09$, and $t = 2$ months

$= 2 \text{ months} \cdot \dfrac{1 \text{ year}}{12 \text{ months}}$

$= \dfrac{2}{12} \text{ year}$

$= \dfrac{1}{6} \text{ year}$

to find the interest owed.

$I = 1,500 \cdot 0.09 \cdot \dfrac{1}{6}$

$= \dfrac{1,500 \cdot 0.09}{6}$

$= \dfrac{135}{6}$

$= 22.5$

At the end of the two months, the student will have to repay $1,500 + $22.50 = $1,522.50.

37. $\dfrac{3 \text{ centimeters}}{7 \text{ centimeters}} = \dfrac{3}{7}$

39.
$$\frac{2t+1}{4} = \frac{15}{12}$$
$$(2t+1)12 = 4(15)$$
$$12(2t+1) = 4(15)$$
$$12 \cdot 2t + 12 \cdot 1 = 4 \cdot 15$$
$$24t + 12 = 60$$
$$24t = 48$$
$$t = 2$$

41. If 98% of the parts are to be within specifications, then 98 of every 100 parts are within specifications, and 2 of every 100 parts are expected to be defective. To find the number of defective parts expected in a run of 1,500 pieces, we use the proportion 2 is to 100 as x is to 1,500.
$$\frac{2}{100} = \frac{x}{1,500}$$
$$2 \cdot 1,500 = 100 \cdot x$$
$$3,000 = 100x$$
$$30 = x$$
30 parts are expected to be defective.

43. $168 \text{ in.} = 168 \text{ in.} \cdot \dfrac{1 \text{ ft}}{12 \text{ in.}}$
$$= \frac{168}{12} \text{ ft}$$
$$= 14 \text{ ft}$$

45. $3 \text{ mi} = 3 \text{ mi} \cdot \dfrac{5,280 \text{ ft}}{1 \text{ mi}} \cdot \dfrac{1 \text{ yd}}{3 \text{ ft}}$
$$= \frac{3 \cdot 5,280}{3} \text{ yd}$$
$$= 5,280 \text{ yd}$$

47. $25 \text{ c} = 25 \text{ c} \cdot \dfrac{8 \text{ fl oz}}{1 \text{ c}}$
$$= 25 \cdot 8 \text{ fl oz}$$
$$= 200 \text{ fl oz}$$

49. $738 \text{ min} = 738 \text{ min} \cdot \dfrac{1 \text{ hr}}{60 \text{ min}}$
$$= \frac{738}{60} \text{ hr}$$
$$= 12.3 \text{ hr}$$

51. $20 \text{ dm} = 20 \text{ dm} \cdot \dfrac{1 \text{ m}}{10 \text{ dm}} \cdot \dfrac{1,000 \text{ mm}}{1 \text{ m}}$
$$= \frac{20 \cdot 1,000}{10} \text{ mm}$$
$$= 2,000 \text{ mm}$$

53. $5 \text{ kg} = 5 \text{ kg} \cdot \dfrac{1,000 \text{ g}}{1 \text{ kg}} \cdot \dfrac{100 \text{ cg}}{1 \text{ g}}$

$= 5 \cdot 1,000 \cdot 100 \text{ cg}$

$= 500,000 \text{ cg}$

55. $20 \text{ g} = 20 \text{ g} \cdot \dfrac{100 \text{ cg}}{1 \text{ g}}$

$= 20 \cdot 100 \text{ cg}$

$= 2,000 \text{ cg}$

57. $5 \text{ kL} = 5 \text{ kL} \cdot \dfrac{1,000 \text{ L}}{1 \text{ kL}}$

$= 5 \cdot 1,000 \text{ L}$

$= 5,000 \text{ L}$

59. $66.04 \text{ cm} = 66.04(0.3937 \text{ in.})$

$\approx 26 \text{ in.}$

61. $600 \text{ oz} = 600(28.35 \text{ g})$

$\approx 17,010 \text{ g}$

63. $F = \dfrac{9}{5}C + 32$

$F = \dfrac{9}{5}(75) + 32$

$= 9 \cdot 15 + 32$

$= 135 + 32$

$= 167°\text{F}$

65. A right angle is 90 degrees.

67. The measures of supplementary angles sum to 180°. Let $x°$ be the measure of the supplement of an angle of 105°.

$x + 105 = 180$

$x = 75$

The supplement of an angle of 105° is an angle of 75°.

69. $m(\angle 1) = 50°$ since $\angle 1$ is the supplement of the 130° angle.

71. $m(\angle 3) = 50°$ since $\angle 3$ and the 130° angle are interior angles on the same side of the transversal of parallel lines, so they are supplementary angles.

73. Since $m\left(\overline{AC}\right) = m\left(\overline{BC}\right)$, $\triangle ABC$ is an isosceles triangle and $\angle A$ has the same measure as $\angle CBA$.

Thus, $m(\angle A) = 75°$. Since $\angle 1$ and $\angle A$ are corresponding angles formed by a transversal of parallel lines,

$m(\angle 1) = m(\angle A) = 75°$.

75. $m(\angle 2) = 105°$ since $\angle 2$ and $\angle 1$ are supplementary angles and $m(\angle 1) = 75°$.

77. $S = (n - 2)180°$
$S = (5 - 2)180°$
$= (3)180°$
$= 540°$

79. 17 animals is to 5 acres as x animals is to 275 acres.

$\dfrac{17}{5} = \dfrac{x}{275}$

$17 \cdot 275 = 5 \cdot x$

$4,675 = 5x$

$935 = x$

There are approximately 935 deer in the forest.

81. For a rectangle, $P = 2l + 2w$ and $A = lw$.

$$P = 2(9) + 2(12)$$
$$= 18 + 24$$
$$= 42$$
$$A = (9)(12)$$
$$= 108$$

The perimeter is 42 m and the area is 108 m².

83. For a trapezoid, $A = \dfrac{1}{2}h(b_1 + b_2)$.

$$A = \frac{1}{2}(7)(12 + 14)$$

$$= \frac{1}{2}(7)(26)$$

$$= 91$$

The area is 91 in.².

85. For a rectangular solid, $V = lwh$.

$$V = (5)(6)(7)$$
$$= 210$$

The volume is 210 m³.

87. For a cone, $V = \dfrac{1}{3}\pi r^2 h$. The diameter of the base is 8 m, so the radius is

$\dfrac{1}{2}(8 \text{ m}) = 4 \text{ m}$.

$$V = \frac{1}{3}\pi(4)^2(9)$$

$$= \frac{1}{3}\pi(16)(9)$$

$$= 48\pi$$

$$\approx 150.80$$

The volume is about 150.80 m³.